CAMBRIDGE TRACTS IN MATHEMATICS
General Editors
B. BOLLOBAS, H. HALBERSTAM, & C. T. C. WALL

---

92 *Heat kernels and spectral theory*

E. B. DAVIES

Professor of Mathematics
King's College, London

Heat kernels and spectral theory

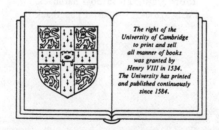

CAMBRIDGE UNIVERSITY PRESS

Cambridge

New York   Port Chester

Melbourne   Sydney

CAMBRIDGE UNIVERSITY PRESS
Cambridge, New York, Melbourne, Madrid, Cape Town, Singapore, São Paulo

Cambridge University Press
The Edinburgh Building, Cambridge CB2 8RU, UK

Published in the United States of America by Cambridge University Press, New York

www.cambridge.org
Information on this title: www.cambridge.org/9780521361361

© Cambridge University Press 1989

This publication is in copyright. Subject to statutory exception
and to the provisions of relevant collective licensing agreements,
no reproduction of any part may take place without the written
permission of Cambridge University Press.

First published 1989
First paperback edition 1990
Re-issued in this digitally printed version 2007

A catalogue record for this publication is available from the British Library

Library of Congress Cataloguing in Publication data
Davies, E. B. (Edward Brian)
Heat kernels and spectral theory/E. B. Davies.
p.   cm.—(Cambridge tracts in mathematics; 92)
Includes bibliographical references and index.
ISBN 0 521 36136 2
1. Elliptic operators. 2. Heat equation. 3. Spectral theory
(Mathematics) I. Title. II. Series.
QA329.42.D38 1989
515.7′242—dc19   88-11629
CIP

ISBN 978-0-521-36136-1 hardback
ISBN 978-0-521-40997-1 paperback

Contents

Preface	vii
1 Introductory concepts	**1**
1.1 Some classical analysis	1
1.2 Quadratic forms	6
1.3 The Beurling–Deny conditions	12
1.4 Symmetric Markov semigroups	21
1.5 An inequality of Hardy	25
1.6 Compactness and spectrum	35
1.7 Some Sobolev inequalities	40
1.8 Definition of Schrödinger operators	48
1.9 The asymptotic eigenvalue distribution	52
Notes	55
2 Logarithmic Sobolev inequalities	**59**
2.1 Contractivity properties	59
2.2 The fundamental inequalities	63
2.3 Examples and applications	71
2.4 Ultracontractivity, Sobolev and Nash inequalities	75
Notes	80
3 Gaussian bounds on heat kernels	**82**
3.1 Introduction	82
3.2 The upper bound	83
3.3 The lower bound	91
3.4 Functions of an elliptic operator	99
Notes	105
4 Boundary behaviour	**107**
4.1 Introduction	107
4.2 Transference to weighted L^2 spaces	109

4.3 The harmonic oscillator 113
4.4 Rosen's lemma 117
4.5 Schrödinger operators 119
4.6 Elliptic operators on bounded regions 125
4.7 Singular elliptic operators 131
4.8 Potentials with local singularities 138
Notes 144

5 Riemannian manifolds **146**
5.1 Fundamental properties of manifolds 146
5.2 Regularity properties of the heat equation 149
5.3 The parabolic Harnack inequality 157
5.4 Potential theory 164
5.5 Upper bound on the heat kernel 167
5.6 Lower bounds on the heat kernel 173
5.7 Hyperbolic space 176
Notes 183

References 187
Notation index 195
Index 196

Preface

This book is a study of linear, self-adjoint, second order, elliptic differential operators. The goal is to investigate spectral properties and obtain pointwise bounds on eigenfunctions by studying the heat kernels.

There is an enormous literature on heat kernels which stretches back half a century, so it is easy to imagine that the subject has already reached its final form. However, we shall make almost no reference to this literature, and shall rely entirely upon results proved within the last five years using quadratic form techniques and logarithmic Sobolev inequalities. These new techniques have led to radically better global bounds on the heat kernels.

We shall be concerned to obtain pointwise upper and lower bounds on various functions in terms of effectively computable constants. In a number of cases the new methods yield constants which are sharp or at least of the correct order of magnitude. This is in sharp distinction to much of the older theory, where various constants appeared to depend upon the magnitudes of the derivatives of the second order coefficients of the differential operator, although in fact they do not.

Because of our approach we are able to deal simply and naturally with operators in divergence form whose second order coefficients are measurable. Earlier treatments of this problem such as that of Gilbarg and Trudinger have relied heavily upon Moser's Harnack inequality. In spite of its fundamental historical and conceptual importance, we make no mention of Moser's approach. The reason is that the proof is fairly lengthy and the constants it produces bear little relationship to the true values. Our lower bounds depend instead upon the reworking of earlier ideas of Nash by Fabes and Stroock.

In spite of the obvious probabilistic undercurrents we work in a purely analytical setting, and make no use of techniques such as functional integration or stochastic differential equations. One cannot predict the future, but at present one gets the sharpest information about operators in divergence form by analytical techniques, while for operators in nondivergence form (which we do not study) one gets the sharpest information by probabilistic techniques.

We have also avoided the use of the wave equation, which has yielded profound and detailed results, but which appears to require smooth coefficients. The reason for this is that the heat equation can be studied on L^p for all $1 \leqslant p \leqslant \infty$, but the wave equation can only be studied in Sobolev spaces if $p = 2$. Pointwise bounds therefore necessitate the availability of suitable elliptic regularity theorems, if the wave equation is used. Even when this is possible their use leads to some loss of sharpness of the bounds.

Although the theory which we describe could have been carried out on a manifold from the start, we have chosen to develop it on a region in Euclidean space. It is our hope that this will make the material more accessible to those interested in Schrödinger operators or variable coefficient elliptic differential operators. However the techniques are applied to study the heat kernels of Laplace–Beltrami operators on Riemannian manifolds in Chapter 5.

The core of the book is Chapter 2, where we present Gross' theory of logarithmic Sobolev inequalities, as adapted by Davies and Simon to yield ultracontractive bounds. Although the ideas are basically simple, the theory is presented in a very abstract and general setting. We were forced to do this by the many variations of the basic argument which are needed to cope with the applications. While many of the applications can also be worked out by means of Sobolev or Nash inequalities, we believe that the use of logarithmic Sobolev inequalities is more natural. Be that as it may, the use of logarithmic Sobolev inequalities is certainly necessary in order to deal with anharmonic Schrödinger operators in Chapter 4.

In Chapter 3 we use the ultracontractive bounds to obtain Gaussian upper bounds on heat kernels. These bounds are sharp in a certain sense and improve greatly upon what was available a decade ago. As well as being important for their own sake, the upper bounds are used to obtain the lower bounds, thus completing the approach which Nash introduced many years ago.

Chapter 4 is an introduction to the study of the behaviour of heat kernels and eigenfunctions near the boundary of the region on which the elliptic operator acts. Upon examining some examples, one realises that this is an extraordinarily complicated subject, even if the coefficients and boundary are piecewise smooth. The main conclusion of the chapter is that the behaviour of the heat kernel and Green function near the boundary is controlled by the behaviour of the ground state. The corresponding statement for Schrödinger operators, however, depends upon the rate at which the potential increases at infinity, the harmonic oscillator being the borderline case.

In the final chapter, where we study Laplace–Beltrami operators on Riemannian manifolds, we make much less effort to give a self-contained treatment. We study only complete Riemannian manifolds whose Ricci curvature is bounded below, but make no hypothesis of bounded geometry. We present the parabolic Harnack inequality of Li and Yau and also a modification of their approach to the proof of heat kernel bounds for manifolds of non-negative Ricci curvature. By contrast to the case where the Ricci curvature may be negative, the above theory is fairly complete.

At the end of each chapter we have included some historical notes as well as supplementary information. We have made no attempt to cover literature prior to Gross' fundamental paper of 1976, and have taken Chavel (1984), Davies (1980), Gilbarg and Trudinger (1977), and Stein (1970) as standard references. Our task has been made easier by the fact that many of the most important papers are less than five years old, but we wish to apologise to all those whose contributions have not been acknowledged.

I wish to conclude by offering my thanks to the many colleagues who have influenced my thinking in the field, and who have helped me to avoid at least some of the historical and mathematical blunders in early versions of this book. I would particularly like to mention Eric Carlen, Nikolaos Mandouvalos, Barry Simon, Daniel Stroock and Nicholas Varopoulos, with whom I have had many stimulating conversations, and without whose help this book might never have been started.

King's College, London   E. B. Davies

1
Introductory concepts

1.1 Some classical analysis

Let Ω be a locally compact, second countable, Hausdorff topological space and let dx be a Borel measure on Ω. If $1 \leq p < \infty$ then we define the real Banach space $L^p(\Omega, dx)$ to be the vector space of measurable functions $f : \Omega \to \mathbb{R}$ for which the norm

$$\|f\|_p = \left\{ \int_\Omega |f(x)|^p \, dx \right\}^{1/p}$$

is finite. We identify two functions which coincide outside a null set without further comment. The space $L^\infty(\Omega, dx)$ is defined to be the space of functions f for which

$$\|f\|_\infty = \min\{\lambda : \operatorname{meas}\{|f| > \lambda\} = 0\}$$

is finite. It may be shown that

$$\|f\|_\infty = \lim_{p \to +\infty} \|f\|_p$$

whenever $f \in L^p$ for all large enough p.

The spaces L^p are all Banach lattices in the sense that

$$\|f\|_p = \| |f| \|_p$$

and

$$\|g\|_p \leq \|f\|_p$$

whenever

$$-f \leq g \leq f.$$

We shall make much use of the ordering of L^p, and introduce the notation

$$(f \wedge g)(x) = \min\{f(x), g(x)\}$$
$$(f \vee g)(x) = \max\{f(x), g(x)\}$$
$$f_+ = f \vee 0, \quad f_- = (-f) \vee 0$$

so that

$$|f| = f_+ + f_-, \quad f = f_+ - f_-.$$

The complex Banach space $L^p_{\mathbb{C}}$ is the algebraic sum $L^p \oplus iL^p$, and the norm is

$$\|f + ig\|_p = \left\{ c \int_\Omega \int_{-\pi}^{\pi} |f \cos\theta + g \sin\theta|^p \, d\theta \, dx \right\}^{1/p}$$

$$= \left\{ c \int_{-\pi}^{\pi} \|f \cos\theta + g \sin\theta\|_p^p \, d\theta \right\}^{1/p} \quad (1.1.1)$$

where

$$c^{-1} = \int_{-\pi}^{\pi} |\cos\theta|^p \, d\theta. \quad (1.1.2)$$

Every bounded linear operator A on $L^p(\Omega)$ extends uniquely to a bounded linear operator $A_{\mathbb{C}}$ on $L^p_{\mathbb{C}}(\Omega)$ which has the same norm and satisfies

$$A_{\mathbb{C}}(\bar{f}) = (A_{\mathbb{C}} f)^-$$

for all $f \in L^p_{\mathbb{C}}(\Omega)$. Although we shall deal mainly with the real Banach spaces, we will use the complex spaces when discussing questions involving analyticity or spectral theory, and will not distinguish between the two notationally.

We will use the following standard results from L^p theory frequently, and without comment.

1.1.1 Duality

If $1 \leq p < \infty$ then $(L^p)^*$ can be identified with L^q where

$$1/p + 1/q = 1.$$

1.1.2 Hölder's inequality

Let $f \in L^p$ and $g \in L^q$ where $1 \leq p \leq \infty$ and $1 \leq q \leq \infty$. If

$$1/p + 1/q = 1/r$$

and $1 \leq r \leq \infty$ then $fg \in L^r$ and

$$\|fg\|_r \leq \|f\|_p \|g\|_q.$$

1.1.3 The Hausdorff–Young inequality

Let dx be the Lebesgue measure on $\Omega = \mathbb{R}^N$. We define the Fourier transform $\mathscr{F}f$ of the function $f \in L^1 \cap L^2$ by

$$\mathscr{F}f(y) = (2\pi)^{-N/2} \int_{\mathbb{R}^N} f(x) e^{-ix \cdot y} \, dx.$$

If $1 \leq p \leq 2$ and $p^{-1} + q^{-1} = 1$ then
$$\|\mathscr{F}f\|_q \leq (2\pi)^{N/2 - N/p} \|f\|_p$$
for all $f \in L^1 \cap L^2$. Therefore $\mathscr{F}$ may be extended to a bounded linear operator from L^p to L^q for such p.

1.1.4 Young's inequality

Let dx be the Lebesgue measure on $\Omega = \mathbb{R}^N$ and let $f \in L^p$, $g \in L^q$ where $1 \leq p \leq \infty$ and $1 \leq q \leq \infty$. If $1 \leq r \leq \infty$ and
$$1 + 1/r = 1/p + 1/q$$
then the convolution $f * g$ lies in L^r and
$$\|f * g\|_r \leq \|f\|_p \|g\|_q.$$

1.1.5 The Riesz–Thorin interpolation theorem

Let $1 \leq p_0, p_1, q_0, q_1 \leq \infty$ and let A be a linear operator from $L^{p_0} \cap L^{p_1}$ to $L^{q_0} + L^{q_1}$ which satisfies
$$\|Af\|_{q_i} \leq M_i \|f\|_{p_i}$$
for all f and $i = 1, 2$. Let $0 < t < 1$ and define p, q by
$$1/p = t/p_1 + (1-t)/p_0, \quad 1/q = t/q_1 + (1-t)/q_0.$$
Then
$$\|Af\|_q \leq M_1^t M_0^{1-t} \|f\|_p$$
for all $f \in L^{p_0} \cap L^{p_1}$. Hence A can be extended to a bounded operator from L^p to L^q with norm at most $M_1^t M_0^{1-t}$.

This is a special case of the following.

1.1.6 The Stein interpolation theorem

Let p_i, q_i, p, q, t be as above and let
$$S = \{z \in \mathbb{C} : 0 \leq \operatorname{Re} z \leq 1\}.$$
Let A_z be linear operators from $L^{p_0} \cap L^{p_1}$ to $L^{q_0} + L^{q_1}$ for all $z \in S$ with the following properties.

(i) $\langle A_z f, g \rangle$ is uniformly bounded and continuous on S and analytic in the interior of S whenever $f \in L^{p_0} \cap L^{p_1}$ and $g \in L^{r_0} \cap L^{r_1}$ where
$$1/q_i + 1/r_i = 1.$$

(ii) $$\|A_{iy} f\|_{q_0} \leq M_0 \|f\|_{p_0}$$
for all $f \in L^{p_0} \cap L^{p_1}$ and $y \in \mathbb{R}$.

(iii) $$\|A_{1+iy}f\|_{q_1} \leq M_1 \|f\|_{p_1}$$

for all $f \in L^{p_0} \cap L^{p_1}$ and $y \in \mathbb{R}$.

Then
$$\|A_t f\|_q \leq M_1^t M_0^{1-t} \|f\|_p$$

for all $f \in L^{p_0} \cap L^{p_1}$. Hence A_t can be extended to a bounded operator from L^p to L^q with norm at most $M_1^t M_0^{1-t}$.

In this theorem the initial domain $L^{p_0} \cap L^{p_1}$ of A_z can be replaced by other domains, such as the set of all functions of the form

$$f = \sum_{r=1}^{n} \alpha_r \chi_{E(r)}$$

where $\alpha_r \in \mathbb{R}$ and $\chi_{E(r)}$ are the characteristic functions of the sets $E(r)$, assumed all to have finite measure.

We now give some standard results and formulae for the Laplace operator acting on $L^2(\mathbb{R}^N)$.

1.1.7 Domain and spectrum of $-\Delta$

The Laplacian $H_0 = -\Delta$ is defined on the space $C_c^\infty(\mathbb{R}^N)$ of infinitely differentiable functions of compact support by

$$H_0 f = -\sum_{r=1}^{N} \partial^2 f / \partial x_r^2$$

and satisfies

$$\langle H_0 f, f \rangle = \int_{\mathbb{R}^N} \sum_{r=1}^{N} |\partial f / \partial x_r|^2 \, dx$$

on that domain, so that $H_0 \geq 0$. It is known that H_0 is essentially self-adjoint on Schwartz space $\mathscr{S}$, or even on $C_c^\infty(\mathbb{R}^N)$. The spectral resolution of H_0 is achieved using the Fourier transform. Explicitly we have

$$(H_0 f)^\wedge(y) = y^2 \hat{f}(y)$$

on the maximal domain for which the RHS lies in L^2. It follows that the spectrum of H_0 is $[0, \infty)$ and is purely absolutely continuous.

1.1.8 The heat kernel and Green function

By the use of Fourier transforms one sees that

$$e^{-H_0 t} f = K_t * f$$

for all $t > 0$ and $f \in L^2(\mathbb{R}^N)$ where

$$K_t(x) = (4\pi t)^{-N/2} e^{-x^2/4t}.$$

Using the formula

$$(H_0 + \lambda)^{-1} = \int_0^\infty e^{-H_0 t} e^{-\lambda t} dt$$

one deduces that if $\operatorname{Re} \lambda > 0$ one has

$$(H_0 + \lambda)^{-1} = G_\lambda * f$$

where

$$G_\lambda(x) = \int_0^\infty (4\pi t)^{-N/2} e^{-x^2/4t} e^{-\lambda t} dt.$$

The kernel G_λ is strictly positive and becomes infinite as $x \to 0$. It is dominated pointwise by the kernel G_0 of the unbounded operator H_0^{-1}, which is given by

$$\begin{aligned} G_0(x) &= \int_0^\infty (4\pi t)^{-N/2} e^{-x^2/4t} dt \\ &= c_N |x|^{-(N-2)} \end{aligned}$$

provided $N > 2$.

1.1.9 Other functions of the Laplacian

If $F \in L^1(\mathbb{R}^N)$ is a spherically symmetric function then there is a bounded continuous function ϕ on $\mathbb{R}$ such that

$$F * f = \phi(H_0) f$$

for all $f \in L^2$, where $\phi(H_0)$ is defined using the spectral calculus. F and ϕ are related explicitly by

$$\phi(y^2) = \int_{\mathbb{R}^N} F(x) e^{-ix \cdot y} dx.$$

This relationship can be extended to other classes of F and ϕ without difficulty.

1.1.10 Minimax

If H is a non-negative self-adjoint operator on a Hilbert space then the spectrum of H is real, and the bottom of the spectrum is given by

$$\begin{aligned} \alpha &= \inf\{\langle Hf, f\rangle : f \in \operatorname{Dom}(H) \text{ and } \|f\| = 1\} \\ &= \inf\{\|H^{\frac{1}{2}} f\|^2 : f \in \operatorname{Dom}(H^{\frac{1}{2}}) \text{ and } \|f\| = 1\}. \end{aligned}$$

A point λ of the spectrum is said to be in the discrete spectrum if it is isolated *and* of finite multiplicity; otherwise it is said to be in the essential spectrum.

If L is a finite dimensional subspace of $\mathrm{Dom}(H^{\frac{1}{2}})$ we define

$$\lambda_H(L) = \sup\{\|H^{\frac{1}{2}}f\|^2 : f \in L \text{ and } \|f\| = 1\}.$$

We then put

$$E_n = \inf\{\lambda_H(L) : \dim L = n+1\}$$

so that $E_0 = \alpha$ and E_n is an increasing sequence. The minimax theorem gives the following information. The least upper bound of $\{E_n\}$ equals the bottom β of the essential spectrum. The sequence $\{E_n\}$, omitting all values equal to β if there are such, coincides with the discrete spectrum of H in the interval $[\alpha, \beta)$, and each eigenvalue is repeated according to its multiplicity.

1.1.11 Comparison of spectra

If H, K are two self-adjoint operators and $H \leq K$ then it is immediate from the definition that

$$\lambda_H(L) \leq \lambda_K(L)$$

for all finite-dimensional subspaces L of $\mathrm{Dom}(H^{\frac{1}{2}})$. It follows that

$$E_n(H) \leq E_n(K)$$

for all n. The bottoms of the essential spectra of the operators are related by

$$\beta(H) \leq \beta(K).$$

1.1.12 The regularised distance function

If Ω is an open set in $\mathbb{R}^N$ then the distance function d on $\mathbb{R}^N$ defined by

$$d(x) = \min\{|x-y| : y \notin \Omega\}$$

is Lipschitz in the sense that

$$|d(x) - d(y)| \leq |x-y|$$

but need not be smooth. A theorem of Whitney states that there exists a constant $c > 0$ and a C^∞ function $\tilde{d}: \Omega \to (0, \infty)$ such that

$$c^{-1}d(x) \leq \tilde{d}(x) \leq cd(x),$$
$$|\nabla \tilde{d}(x)| \leq c,$$
$$|\Delta \tilde{d}(x)| \leq c\tilde{d}(x)^{-1}$$

for all $x \in \Omega$.

1.2 Quadratic forms

We shall make extensive use of the theory of quadratic forms, and summarise some of the important results.

If $\mathscr{D}$ is a linear subspace of a Hilbert space $\mathscr{H}$ a quadratic form on $\mathscr{D}$ is defined to be a map $Q':\mathscr{D} \times \mathscr{D} \to \mathbb{C}$ such that

(i) $\qquad Q'(\alpha f + \beta g, h) = \alpha Q'(f, h) + \beta Q'(g, h)$
(ii) $\qquad Q'(h, \alpha f + \beta g) = \bar{\alpha} Q'(h, f) + \bar{\beta} Q'(h, g)$
(iii) $\qquad Q'(f, g) = \overline{Q'(g, f)}$

for all $f, g, h \in \mathscr{D}$ and $\alpha, \beta \in \mathbb{C}$. We shall not distinguish between Q' and the map Q from $\mathscr{H}$ to $(-\infty, +\infty]$ defined by

$$Q(f) = \begin{cases} Q'(f, f) & \text{if } f \in \mathscr{D} \\ +\infty & \text{otherwise.} \end{cases}$$

We say that Q is bounded below if there exists $c \in \mathbb{R}$ such that

$$Q(f) \geq c \|f\|^2 \qquad (1.2.1)$$

for all $f \in \mathscr{H}$, and that Q is non-negative if one may take $c = 0$ in (1.2.1). If Q is bounded below, we say that it is closed if for all sequences $f_n \in \mathscr{D}$ such that

$$\lim_{n \to \infty} \|f_n - f\| = 0, \quad \lim_{m,n \to \infty} Q(f_m - f_n) = 0$$

it follows that $f \in \mathscr{D}$ and

$$\lim_{n \to \infty} Q(f_n - f) = 0.$$

If $H \geq 0$ is a self-adjoint operator on a closed linear subspace $\mathscr{L}$ of $\mathscr{H}$ then we define its form by

$$Q(f) = \begin{cases} \langle H^{\frac{1}{2}} f, H^{\frac{1}{2}} f \rangle & \text{if } f \in \mathrm{Dom}(H^{\frac{1}{2}}) \\ +\infty & \text{otherwise} \end{cases}$$

and we write $\mathrm{Dom}(Q)$ or $\mathrm{Quad}(H)$ for the domain of $H^{\frac{1}{2}}$ as convenient. We quote the following fundamental result.

Theorem 1.2.1. *If Q is a non-negative form on $\mathscr{H}$ with domain $\mathscr{D}$, then the following conditions are equivalent.*
  (i) *Q is the form of a self-adjoint operator $H \geq 0$ on $\mathscr{L} = \bar{\mathscr{D}}$.*
  (ii) *Q is closed.*
  (iii) *Q is lower semicontinuous as a function from $\mathscr{H}$ to $[0, +\infty]$.*

If Q is a non-negative form then its domain is an inner product space with inner product

$$\langle f, g \rangle_Q = \langle f, g \rangle + Q(f, g) \qquad (1.2.2)$$

and is complete if and only if Q is closed. A subspace of $\mathrm{Dom}(Q)$ is said to be a form core of Q if it is dense with respect to the norm $\| \quad \|_Q$ associated to (1.2.2). Q is said to be closable if it has a closed extension and the closure $\bar{Q}$ is then the least closed extension. It is obvious from Theorem 1.2.1 (iii) that the sum of two closed forms is closed. We shall need two limit theorems.

Theorem 1.2.2. *Let H_n be an increasing sequence of non-negative self-adjoint operators on $\mathcal{H}$. Let Q_n be the associated forms and define Q on $\mathcal{H}$ by*

$$Q(f) = \lim_{n\to\infty} Q_n(f)$$

so that

$$\mathrm{Dom}(Q) \subseteq \bigcap_n \mathrm{Quad}(H_n).$$

Then there exists a self-adjoint operator $H \geq 0$ on the closure $\mathcal{H}_0$ of $\mathrm{Dom}(Q)$ such that

$$\langle H^{\frac{1}{2}}f, H^{\frac{1}{2}}f \rangle = \lim_{n\to\infty} \langle H_n^{\frac{1}{2}}f, H_n^{\frac{1}{2}}f \rangle$$

for all $f \in \mathrm{Dom}(Q)$.

The operator H will be called the form limit of H_n, and we shall mostly use this theorem with $\mathcal{H}_0 = \mathcal{H}$.

Theorem 1.2.3. *Let H_n and H be non-negative self-adjoint operators on $\mathcal{H}$ such that*

$$H_n \geq H_{n+1} \geq H$$

for all n. Suppose also that the associated forms satisfy

$$\lim_{n\to\infty} Q_n(f) = Q(f)$$

for all f in a form core of Q. Then H_n converges to H in the strong resolvent sense.

We shall use the following notation frequently. If $\mathcal{B}(\Omega)$ is a Banach space of functions on a region (open connected set) Ω in $\mathbb{R}^N$, such as $L^p(\Omega)$, then we write $\mathcal{B}_c(\Omega)$ for the set of f in $\mathcal{B}(\Omega)$ which have compact support, and $\mathcal{B}_0(\Omega)$ for the completion of $\mathcal{B}_c(\Omega)$ in $\mathcal{B}(\Omega)$. We also write $\mathcal{B}_{\mathrm{loc}}(\Omega)$ for the class of functions on Ω which coincide on any compact subset K of Ω with some (K-dependent) element of $\mathcal{B}(\Omega)$.

We shall make particular use of the Sobolev space

$$W^{1,p}(\Omega) = \{f \in L^p(\Omega) : \nabla f \in L^p(\Omega)\}$$

where ∇f is calculated in the weak sense, and we always assume $1 \leq p \leq \infty$. It is easy to see that $W^{1,p}(\Omega)$ is a Banach space for an appropriate norm. If $1 \leq p < \infty$ we shall take the norm to be

$$\|f\| = (\|f\|_p^p + \|\nabla f\|_p^p)^{1/p} \qquad (1.2.3)$$

so that $W^{1,2}(\Omega)$ is a Hilbert space.

Lemma 1.2.4. *If $1 \leq p < \infty$ then $W_0^{1,p}(\Omega)$ is the closure in $W^{1,p}(\Omega)$ of $C_c^\infty(\Omega)$.*

Proof. This depends upon a standard mollifier argument which we shall use frequently. Let ϕ be a non-negative C^∞ function on $\mathbb{R}^N$ with support in the unit ball and integral equal to 1. If $f \in W_c^{1,p}(\Omega)$ and $\varepsilon > 0$ put $f_\varepsilon = f * \phi_\varepsilon$ where

$$\phi_\varepsilon(x) = \varepsilon^{-N}\phi(x/\varepsilon).$$

It is easy to see that $f_\varepsilon \in C_c^\infty(\Omega)$ for small enough ε and that f_ε converges to f in the norm (1.2.3) as $\varepsilon \to 0$.

We shall be concerned with self-adjoint second order partial differential operators on $L^2(\Omega)$ where Ω is a region in $\mathbb{R}^N$. These operators will be constructed starting from quadratic forms on $C_c^\infty(\Omega)$ of the type

$$Q(f) = \int_\Omega \sum_{i,j} a_{ij}(x) \frac{\partial f}{\partial x_i} \frac{\partial \bar{f}}{\partial x_j} dx \tag{1.2.4}$$

where $a(x)$ is a locally integrable function on Ω with values in the non-negative real symmetric matrices. There are two standard conditions under which Q may be proved to be closable.

Theorem 1.2.5. *If $a \in W_{\text{loc}}^{1,2}(\Omega)$ then Q is closable and the self-adjoint operator H associated with the closure is an extension of the operator*

$$Lf = -\sum_{i,j} \frac{\partial}{\partial x_i}\left(a_{ij}\frac{\partial f}{\partial x_j}\right) \tag{1.2.5}$$

defined on $C_c^\infty(\Omega)$.

Proof. The condition on a implies that L maps C_c^∞ into $L^2(\Omega)$ and integration by parts yields the identity

$$\langle Lf, f \rangle = Q(f)$$

for all $f \in C_c^\infty$, so $L \geq 0$. We may now apply Davies (1980), Theorem 4.14.

Theorem 1.2.6. *If $a \in L_{\text{loc}}^1(\Omega)$ and $a(x) \geq \lambda(x) 1$ for all $x \in \Omega$ where λ is a strictly positive continuous function, then Q is closable on $C_c^\infty(\Omega)$. If f lies in the domain of the closure then (1.2.4) is valid where $\nabla f \in L_{\text{loc}}^2$ is interpreted in the weak sense.*

Proof. Let $\mathscr{D} \subseteq L^2(\Omega)$ consist of all functions which lie in $W_{\text{loc}}^{1,2}(\Omega)$ and for which the integral

$$\tilde{Q}(f) = \int_\Omega \sum_{i,j} a_{ij}(x) \frac{\partial f}{\partial x_i} \frac{\partial \bar{f}}{\partial x_j} dx$$

is finite. A routine calculation shows that $\mathscr{D}$ is complete for the norm

$$|||f||| = (\|f\|_2^2 + \tilde{Q}(f))^{\frac{1}{2}}.$$

Therefore the form $\tilde{Q}$ is complete on $\mathscr{D}$. But Q is a restriction of $\tilde{Q}$ so Q is closable.

If $H \geqslant 0$ is the self-adjoint operator on $L^2(\Omega)$ associated with the closed form Q obtained in Theorem 1.2.5 or 1.2.6, then we say that H satisfies Dirichlet boundary conditions (in the generalised sense). Although one has

$$Hf = -\sum_{i,j} a_{ij}\frac{\partial^2 f}{\partial x_i \partial x_j} - \sum_{i,j} \frac{\partial a_{ij}}{\partial x_i}\frac{\partial f}{\partial x_j} \qquad (1.2.6)$$

in a formal sense, we shall be particularly concerned not to assume that $a_{ij}(x)$ is differentiable. If, however, Ω has smooth boundary and $a_{ij}(x)$ are smooth then the operator H is indeed given by (1.2.6) with Dirichlet boundary conditions in the classical sense.

Henceforth the phrase 'H is an elliptic operator' will mean that H is constructed by the procedure of Theorem 1.2.6 and that

$$\lambda(x)1 \leqslant a(x) \leqslant \mu(x)1 \qquad (1.2.7)$$

where λ and μ are two positive continuous functions on Ω. We will say that H is strictly elliptic if we can take λ to be a positive constant, and uniformly elliptic if we can take λ and μ to be positive constants. For uniformly elliptic operators we have

$$\text{Dom}(\bar{Q}) = \text{Quad}(H) = W_0^{1,2}(\Omega).$$

Theorem 1.2.7. *If H is uniformly elliptic on $L^2(\Omega)$ then $f \in \text{Dom}(H)$ if $f \in W_0^{1,2}(\Omega)$ and there exists $g \in L^2$ such that $Lf = g$ in the sense that*

$$\int_\Omega \sum_{i,j} a_{ij}\frac{\partial f}{\partial x_i}\frac{\partial \bar{u}}{\partial x_j} dx = \int g\bar{u} dx$$

for all $u \in C_c^\infty$. We then have $Hf = g$.

Proof. In abstract terms this is the assertion that $f \in \text{Dom}(H)$ if and only if $f \in \text{Dom}(H^{\frac{1}{2}})$ and there exists $g \in L^2$ such that

$$\langle H^{\frac{1}{2}}f, H^{\frac{1}{2}}u \rangle = \langle g, u \rangle \qquad (1.2.8)$$

for all u in the core C_c^∞ of $\text{Dom}(H^{\frac{1}{2}})$. By taking limits we may assume that (1.2.8) holds for all $u \in \text{Dom}(H^{\frac{1}{2}})$ and then appeal to the spectral theorem as in Davies (1980), Theorem 4.12.

We next give another condition for a non-negative quadratic form to be closable.

Theorem 1.2.8. *Let $\mathscr{D}$ be a dense linear subspace of $\mathscr{H}$ and let*

$$Q(f) = \langle Hf, f \rangle \geqslant 0$$

for all $f \in \mathcal{D}$ where H is a symmetric linear operator with domain $\mathcal{D}$. Then Q is closable and its closure is associated with a self-adjoint operator $K \geq 0$ on $\mathcal{H}$, which is an extension of H, and called its Friedrichs extension.

We shall need the next lemma to deal with elliptic operators satisfying Neumann boundary conditions.

Lemma 1.2.9. *If Ω is a region in $\mathbb{R}^N$ and $1 \leq p < \infty$ then $C^\infty(\Omega) \cap W^{1,p}(\Omega)$ is dense in $W^{1,p}(\Omega)$ for the norm*

$$\|\|f\|\| = \{\|f\|_p^p + \|\nabla f\|_p^p\}^{1/p}.$$

If $f \in W^{1,p}(\Omega)$ then $g = |f| \in W^{1,p}(\Omega)$ and $\|\|g\|\| \leq \|\|f\|\|$, with equality if f is real.

Proof. For the first statement we refer to Adams (1975) p. 51. If $f \in W^{1,p}(\Omega)$ and $f_n \in C^\infty(\Omega) \cap W^{1,p}(\Omega)$ converge in norm to f then $g_n = |f_n|$ is a Cauchy sequence in $W^{1,p}(\Omega)$ which converges in L^p norm to g. Therefore g_n converges in the $W^{1,p}(\Omega)$ norm to g. The stated norm bounds follow easily.

The following procedure leads to a self-adjoint operator which can often be identified as that associated with Neumann boundary conditions in the classical sense.

Theorem 1.2.10. *If*

$$0 < \lambda 1 \leq a(x) \leq \mu 1 < \infty$$

for all $x \in \Omega$ then the form Q of (1.2.4) is closed on $W^{1,2}(\Omega)$. If also $a \in W^{1,2}_{\text{loc}}(\Omega)$ then the corresponding operator H is given by (1.2.5) interpreted in the weak sense on the domain of H.

Proof. If Q_0 is given on $W^{1,2}(\Omega)$ by

$$Q_0(f) = \int_\Omega |\nabla f|^2 \, dx$$

then Q_0 is closed because $W^{1,2}(\Omega)$ is complete, and Q is then closed because

$$\lambda Q_0 \leq Q \leq \mu Q_0.$$

If $f \in C_c^\infty$ and $g \in W^{1,2}(\Omega)$ then

$$Q(f,g) = \int \sum_{i,j} a_{ij} \frac{\partial f}{\partial x_i} \frac{\partial \bar{g}}{\partial x_j} \, dx$$

$$= -\int f \sum_{i,j} \frac{\partial}{\partial x_i} \left(a_{ij} \frac{\partial \bar{g}}{\partial x_j} \right) dx$$

$$= \langle f, Lg \rangle$$

where L is given by (1.2.5) interpreted in the weak sense. If also $g \in \text{Dom}(H)$ then

$$Q(f,g) = \langle f, Hg \rangle$$

for all $f \in C_c^\infty \subseteq \text{Dom}(Q) = \text{Quad}(H)$, so $Hg = Lg$.

1.3 The Beurling–Deny conditions

We shall need some theorems of Beurling and Deny. Although these are well known they are so fundamental to our later work that we include complete proofs. We start by presenting the theory at an abstract level. We apply it to second order, elliptic operators satisfying Dirichlet or Neumann boundary conditions on a region Ω in $\mathbb{R}^N$ in Theorems 1.3.5 and 1.3.9 respectively. Finally in Theorem 1.3.10 we give an illustration of how the conditions may be applied to certain second order difference operators on discrete sets, which correspond to Poisson jump processes.

We assume that Ω is a locally compact, second countable, Hausdorff space and that dx is a Borel measure on Ω, but comment that even more general cases can be considered below.

Lemma 1.3.1. *Let C be a cone in a real Hilbert space $\mathcal{H}$ and suppose that for all $f \in \mathcal{H}$ there exists $\tilde{f} \in \mathcal{H}$ such that $\|\tilde{f}\| \leq \|f\|$ and*

$$\langle \tilde{f}, c \rangle \geq |\langle f, c \rangle|$$

for all $c \in C$. Then $f = \tilde{f}$ for all $f \in C$.

Proof. If $f \in C$ then $\|\tilde{f}\| \leq \|f\|$ and

$$\|f\|^2 = |\langle f, f \rangle| \leq \langle \tilde{f}, f \rangle,$$

so $f = \tilde{f}$.

Before proving the following theorem we comment that (1.3.2) below may also be written imprecisely as

$$H(|u|) \leq \frac{u}{|u|} H(u), \tag{1.3.1}$$

which in applications may sometimes be interpreted as an inequality between distributions.

Theorem 1.3.2. *Let $H \geq 0$ be a real self-adjoint operator on $\mathcal{H} = L^2(\Omega, dx)$. Then the following are equivalent:*

(i) $u \in \text{Quad}(H)$ implies $|u| \in \text{Quad}(H)$. Also $u \in \text{Dom}(H)$, $f \in \text{Quad}(H)$ and $f \geq 0$

imply
$$\langle H^{\frac{1}{2}}f, H^{\frac{1}{2}}|u|\rangle \leqslant \langle f, \frac{u}{|u|}(Hu)\rangle. \quad (1.3.2)$$

(ii) $u \in \text{Quad}(H)$ implies $|u| \in \text{Quad}(H)$ and
$$\langle H^{\frac{1}{2}}|u|, H^{\frac{1}{2}}|u|\rangle \leqslant \langle H^{\frac{1}{2}}u, H^{\frac{1}{2}}u\rangle. \quad (1.3.3)$$

(iii) $(H+\alpha)^{-1}$ is positivity-preserving for all $\alpha > 0$.
(iv) e^{-Ht} is positivity-preserving for all $t \geqslant 0$.

Proof. (i)$\Rightarrow$(ii) If $u \in \text{Dom}(H)$ then (1.3.3) follows from (1.3.2) by putting $f = |u|$. If $u \in \text{Quad}(H)$ then there exists a sequence $u_n \in \text{Dom}(H)$ such that
$$\lim_{n \to \infty} u_n = u, \quad \lim_{n \to \infty} \langle H^{\frac{1}{2}}u_n, H^{\frac{1}{2}}u_n\rangle = \langle H^{\frac{1}{2}}u, H^{\frac{1}{2}}u\rangle$$
by the spectral theorem. Since $\lim_{n \to \infty} |u_n| = |u|$, (1.3.3) follows by the lower semicontinuity of the quadratic form.

(ii)$\Rightarrow$(iii) We make $\mathcal{K} = \text{Quad}(H)$ into a Hilbert space by defining
$$\langle f, g\rangle_1 = \langle H^{\frac{1}{2}}f, H^{\frac{1}{2}}g\rangle + \alpha\langle f, g\rangle$$
where $\alpha > 0$. The injection $J: \mathcal{K} \to \mathcal{H}$ is bounded and
$$J^*f = (H+\alpha)^{-1}f$$
for all $f \in \mathcal{H}$. We define the cone $C \subseteq \mathcal{K}$ by
$$C = \{J^*f : 0 \leqslant f \in \mathcal{H}\}.$$
If $f \in \mathcal{K}$ and $c \in C$ then $c = J^*g$ for some $g \geqslant 0$ and
$$\langle |f|, c\rangle_1 = \langle |f|, J^*g\rangle_1 = \langle J|f|, g\rangle$$
$$= \langle |f|, g\rangle \geqslant |\langle f, g\rangle| = |\langle f, J^*g\rangle_1| = |\langle f, c\rangle_1|.$$
Moreover if $f \in \mathcal{K}$ then (1.3.3) implies $|f| \in \mathcal{K}$ and
$$\||f|\|_1^2 = \langle H^{\frac{1}{2}}|f|, H^{\frac{1}{2}}|f|\rangle + \alpha\||f|\|^2$$
$$\leqslant \langle H^{\frac{1}{2}}f, H^{\frac{1}{2}}f\rangle + \alpha\|f\|^2$$
$$= \|f\|_1^2.$$
Using Lemma 1.3.1 with $\tilde{f} = |f|$ we conclude that if $0 \leqslant f \in \mathcal{H}$ then
$$|J^*f| = J^*f$$
so
$$J^*f \geqslant 0.$$
Therefore $(H+\alpha)^{-1}$ is positivity-preserving.

(iii)$\Rightarrow$(iv) Since e^{-Ht} equals
$$s-\lim_{n \to \infty}\left\{\frac{t}{n}\left(\frac{n}{t}+H\right)\right\}^{-n}$$

it is a limit of positivity-preserving operators and hence is itself positivity-preserving.

(iv)⇒(i) If $f \in \mathrm{Quad}(H)$ then by positivity

$$\langle e^{-Ht}f, f \rangle \leq \langle e^{-Ht}|f|, |f| \rangle,$$

so

$$\langle t^{-1}(1 - e^{-Ht})f, f \rangle \geq \langle t^{-1}(1 - e^{-Ht})|f|, |f| \rangle.$$

Letting $t \to 0$ we obtain

$$\limsup_{t \to 0} \langle t^{-1}(1 - e^{-Ht})|f|, |f| \rangle \leq \langle H^{\frac{1}{2}}f, H^{\frac{1}{2}}f \rangle,$$

which by the spectral theorem implies $|f| \in \mathrm{Quad}(H)$ and

$$\langle H^{\frac{1}{2}}|f|, H^{\frac{1}{2}}|f| \rangle \leq \langle H^{\frac{1}{2}}f, H^{\frac{1}{2}}f \rangle.$$

If $u \in \mathrm{Dom}(H)$, $f \in \mathrm{Quad}(H)$ and $f \geq 0$ then

$$\left| \frac{u}{|u|}(e^{-Ht}u) \right| = |e^{-Ht}u| \leq e^{-Ht}|u|,$$

so

$$\left\langle f, \frac{u}{|u|}(e^{-Ht}u) \right\rangle \leq \langle f, e^{-Ht}|u| \rangle.$$

This implies that

$$\left\langle f, \frac{u}{|u|}t^{-1}(e^{-Ht} - 1)u \right\rangle \leq \langle f, t^{-1}(e^{-Ht} - 1)|u| \rangle$$

for all $t > 0$, from which (1.3.2) follows by letting $t \to 0$.

Theorem 1.3.3. *Let $H \geq 0$ satisfy the conditions of Theorem 1.3.2. Then the following are equivalent:*

(i) e^{-Ht} *is a contraction on L^∞ for all $t \geq 0$.*
(ii) e^{-Ht} *is a contraction on L^p for all $1 \leq p \leq \infty$ and $t \geq 0$.*
(iii) *Let $f \in \mathrm{Quad}(H)$ and let $g \in L^2$ satisfy*

$$|g(x)| \leq |f(x)|,$$
$$|g(x) - g(y)| \leq |f(x) - f(y)|$$

for all $x, y \in \Omega$. Then $g \in \mathrm{Quad}(H)$ and

$$Q(g) \leq Q(f).$$

(iv) *If $0 \leq f \in \mathrm{Quad}(H)$ then $f \wedge 1$ lies in $\mathrm{Quad}(H)$ and*

$$Q(f \wedge 1) \leq Q(f).$$

Proof. When we say that e^{-Ht} is a contraction on L^p for $p \neq 2$ we mean precisely that e^{-Ht}, which is properly only defined on L^2, maps $L^2 \cap L^p$ into $L^2 \cap L^p$ and can be extended to a contraction on L^p. This extension is unique by density for $1 \leq p < \infty$ and is unique for $p = \infty$ if we impose the extra condition of weak* continuity.

(i) $\Rightarrow$ (ii) By duality we see that e^{-Ht} is also a contraction on L^1 and by the Riesz–Thorin interpolation theorem the same follows on L^p for all $1 \leq p \leq \infty$.

(ii) $\Rightarrow$ (iii) Because of the formula

$$Q(f) = \lim_{t \to 0} \langle t^{-1}(1 - e^{-Ht})f, f \rangle$$

which holds whether or not $f \in \text{Dom}(Q)$, we see that it is sufficient to prove that

$$\langle (1 - e^{-Ht})g, g \rangle \leq \langle (1 - e^{-Ht})f, f \rangle \tag{1.3.4}$$

for all $t > 0$.

Because $f, g \to \langle e^{-Ht}f, g \rangle$ is a positive bilinear form on $C_c(\Omega) \times C_c(\Omega)$, there exists a symmetric Borel measure $\mu_t \geq 0$ on $\Omega \times \Omega$ such that

$$\langle e^{-Ht}f, g \rangle = \int_{\Omega \times \Omega} f(x)\overline{g(y)}\, d\mu_t(x, y)$$

for all $f, g \in C_c$ and then for all $f, g \in L^2(\Omega)$. See Davies (1976) p. 51 for the proof of this measure-theoretic result. Because e^{-Ht} is a contraction on L^1 we have

$$0 \leq \mu_t(E \times \Omega) \leq \int_E dx$$

for all Borel sets E. The Radon–Nikodym theorem now implies that there exists a function $\rho_t: \Omega \to [0, 1]$ such that

$$\int_{\Omega \times \Omega} h(x)\, d\mu_t(x, y) = \int_\Omega \rho_t(x) h(x)\, dx$$

for all h. We deduce that

$$\langle (1 - e^{-Ht})f, f \rangle = \int |f(x)|^2\, dx - \int_{\Omega \times \Omega} f(x)\overline{f(y)}\, d\mu_t(x, y)$$

$$= \int_\Omega \{1 - \rho_t(x)\}|f(x)|^2\, dx$$

$$+ \tfrac{1}{2} \int_{\Omega \times \Omega} |f(x) - f(y)|^2\, d\mu_t(x, y).$$

The inequality (1.3.4) follows directly from this representation.

(iii) $\Rightarrow$ (iv) One merely has to check that $g = f \wedge 1$ satisfies the conditions in (iii).

(iv) $\Rightarrow$ (i) Let $f \in L^2$ satisfy $0 \leq f \leq 1$ and put $g = (1 + sH)^{-1}f$ and

$h = g \wedge 1$ where $0 < s < \infty$. Then $g, h \in \text{Quad}(H)$ and

$$\|(1+sH)^{\frac{1}{2}}(g-h)\|^2 = \langle (1+sH)^{-1}f, f \rangle - 2\langle f, h \rangle + \|(1+sH)^{\frac{1}{2}}h\|^2$$
$$\leq \langle (1+sH)^{-1}f, f \rangle - \|f\|^2 + \|f-h\|^2 + sQ(h)$$
$$\leq \langle (1+sH)^{-1}f, f \rangle - \|f\|^2 + \|f-g\|^2 + sQ(g)$$
$$= 0.$$

Therefore $g = h$, or equivalently

$$0 \leq (1+sH)^{-1}f \leq 1.$$

Using the formula

$$e^{-Ht}f = \lim_{n \to \infty} \{1 + (t/n)H\}^{-n}f$$

we deduce that

$$0 \leq e^{-Ht}f \leq 1$$

for all $0 \leq t < \infty$. This implies that e^{-Ht} is a contraction on L^∞ and L^1.

Many applications of the above theorems depend upon the following lemma.

Lemma 1.3.4. *Suppose $H \geq 0$ has a form core $\mathscr{D}$ such that $f \in \mathscr{D}$ implies $|f| \in \text{Quad}(H)$ and*

$$Q(|f|) \leq Q(f). \tag{1.3.5}$$

Then H satisfies the conditions of Theorem 1.3.2. If in addition $f \in \mathscr{D}$ implies $0 \vee (f \wedge 1) \in \text{Quad}(H)$ and

$$Q(0 \vee (f \wedge 1)) \leq Q(f), \tag{1.3.6}$$

then H satisfies the conditions of Theorem 1.3.3.

Proof. Given $f \in \text{Quad}(H)$, let $f_n \in \mathscr{D}$ satisfy

$$\|f_n - f\| \to 0, \quad Q(f_n - f) \to 0$$

so that $Q(f_n) \to Q(f)$. Then

$$\| |f_n| - |f| \| \to 0$$

and

$$\limsup Q(|f_n|) \leq \limsup Q(f_n)$$
$$= Q(f).$$

By the lower semicontinuity of Q we deduce that $|f| \in \text{Quad}(H)$ and

$$Q(|f|) \leq Q(f).$$

Secondly let $0 \leq f \in \text{Quad}(H)$ and let $f_n \in \mathscr{D}$ satisfy

$$\|f_n - f\| \to 0 \quad \text{and} \quad Q(f_n - f) \to 0,$$

We see that $g_n = 0 \vee (f_n \wedge 1)$ satisfy
$$\|g_n - f \wedge 1\| \to 0$$
and
$$\limsup_{n \to \infty} Q(g_n) \leqslant \limsup_{n \to \infty} Q(f_n)$$
$$= Q(f),$$
so by the lower semicontinuity of Q we deduce that $f \wedge 1 \in \text{Quad}(H)$ and
$$Q(f \wedge 1) \leqslant Q(f).$$

Theorem 1.3.5. *If Ω is a region in $\mathbb{R}^N$ and H is an elliptic operator on $L^2(\Omega)$ satisfying Dirichlet boundary conditions then H satisfies the conditions of Theorems 1.3.2 and 1.3.3.*

Proof. We apply Lemma 1.3.4 with $\mathscr{D}$ equal to $C_c^\infty(\Omega)$. Let ϕ be a C^∞ function from $\mathbb{R}$ to $[0, \infty)$ such that $\phi(0) = 0$ and $|\phi'(x)| \leqslant 1$ for all x and $\phi(x) = |x| - 1$ if $|x| \geqslant 2$. Given $f \in \mathscr{D}$ put
$$f_\varepsilon = \varepsilon \phi(\varepsilon^{-1} f)$$
so that $f_\varepsilon \in C_c^\infty$. Since their support sets satisfy
$$\text{supp}(f_\varepsilon) \subseteq \text{supp}(f)$$
and f_ε converges uniformly to $|f|$ we have
$$\lim_{\varepsilon \to 0} \|f_\varepsilon - |f|\| = 0.$$
Moreover a direct calculation shows that
$$Q(f_\varepsilon) \leqslant Q(f)$$
so the lower semicontinuity of Q implies (1.3.5). The proof of (1.3.6) is entirely similar.

The following technical lemma is often useful.

Lemma 1.3.6. *If H is an elliptic operator on $\Omega \subseteq \mathbb{R}^N$ and $0 \leqslant f \in \text{Quad}(H)$ then there exist $f_n \in W_c^{1,2}(\Omega)$ such that $0 \leqslant f_n \leqslant f$ and $\|f_n - f\|_Q \to 0$ for the norm associated with (1.2.2).*

Proof. By the definition of $\text{Quad}(H)$ there exist $g_n \in C^\infty(\Omega)$ with $\|g_n - f\|_Q \to 0$. Now $\||g_n| - f\|_2 \to 0$ and
$$Q(|g_n|) \leqslant Q(g_n) \to Q(f)$$
so
$$\overline{\lim} \, \||g_n|\|_Q \leqslant \|f\|_Q.$$
Therefore
$$\||g_n| - f\|_Q \to 0.$$

We now put $f_n = |g_n| \wedge f$ so that $0 \leq f_n \leq f$ and $\|f_n - f\|_2 \to 0$. Since
$$|g_n| \wedge f = \tfrac{1}{2}(|g_n| + f) - \tfrac{1}{2}||g_n| - f|$$
and
$$Q(||g_n| - f|) \leq Q(|g_n| - f) \to 0$$
we see that
$$Q(|g_n| \wedge f) \to Q(f)$$
so
$$\|f_n - f\|_Q \to 0$$
as required. The method of construction of f_n shows that they lie in $W^{1,2}_{\text{loc}}(\Omega)$ and have compact supports.

We say that an operator H which satisfies the conditions of Theorem 1.3.2 is local if $0 \leq f, g \in \text{Quad}(H)$ and $f \wedge g = 0$ imply

$$\langle H^{\frac{1}{2}} f, H^{\frac{1}{2}} g \rangle = 0. \tag{1.3.7}$$

Lemma 1.3.7. *Let $H \geq 0$ be a real self-adjoint operator on $L^2(\Omega, dx)$. Then H is local if and only if $|u| \in \text{Quad}(H)$ for all real functions $u \in \text{Quad}(H)$, and*

$$\langle H^{\frac{1}{2}} |u|, H^{\frac{1}{2}} |u| \rangle = \langle H^{\frac{1}{2}} u, H^{\frac{1}{2}} u \rangle. \tag{1.3.8}$$

Proof. This is simply a matter of applying (1.3.7) with $f = u_+$ and $g = u_-$.

Theorem 1.3.8. *If Ω is a region on $\mathbb{R}^N$ and H is an elliptic operator on $L^2(\Omega)$ then H is local in the above sense.*

Proof. If $0 \leq f, g \in \text{Quad}(H)$ and $f \wedge g = 0$ then Lemma 1.3.6 implies that there exist $f_n, g_n \in W^{1,2}_c(\Omega)$ such that $0 \leq f_n \leq f$, $0 \leq g_n \leq g$, and $\|f_n - f\|_Q \to 0$ and $\|g_n - g\|_Q \to 0$. Therefore

$$\langle H^{\frac{1}{2}} f, H^{\frac{1}{2}} g \rangle = \lim_{n \to \infty} \langle H^{\frac{1}{2}} f_n, H^{\frac{1}{2}} g_n \rangle$$

and $f_n \wedge g_n = 0$ for all n. It is therefore sufficient to prove (1.3.7) under the extra assumption that $f, g \in W^{1,2}_c(\Omega)$, which we make from now on.

If $f \wedge g = 0$ and f, g both have support in the compact set K then choosing $\alpha \in \mathbb{R}$ so that

$$0 \leq a(x) \leq \alpha 1$$

for all $x \in K$, we obtain

$$0 \leq \langle H^{\frac{1}{2}} f, H^{\frac{1}{2}} g \rangle$$
$$\leq \alpha \langle (-\Delta)^{\frac{1}{2}} f, (-\Delta)^{\frac{1}{2}} g \rangle$$
$$= \alpha \sum_{r=1}^{n} \langle D_r f, D_r g \rangle$$

where $D_r = \partial/\partial x_r$. If we can show that
$$\{x: D_r f(x) \neq 0\} \subseteq \{x: f(x) \neq 0\} \quad (1.3.9)$$
up to a null set, then it will follow from $f \wedge g = 0$ that
$$\langle D_r f, D_r g \rangle = 0$$
for all r, which will complete the proof.

To prove (1.3.9) let $0 \leq h \in L^2(\Omega)$ be any function such that
$$\{x: f(x) \neq 0\} \cap \{h: h(x) \neq 0\} = \varnothing \quad (1.3.10)$$
and consider the function
$$\phi(t) = \langle \exp(D_r t) f, h \rangle.$$
Since $f \in W_c^{1,2}(\Omega) \subseteq \text{Dom}(D_r)$, $\phi(t)$ is differentiable with $\phi(t) \geq 0$ for all t and $\phi(0) = 0$. Therefore
$$0 = \phi'(0) = \langle D_r f, h \rangle.$$
But $h \geq 0$ is arbitrary subject to (1.3.10) so (1.3.9) follows.

We now turn to some other applications of the Beurling–Deny conditions.

Theorem 1.3.9. *If H is a uniformly elliptic operator on $L^2(\Omega)$ satisfying Neumann boundary conditions in the sense of Theorem 1.2.10, then H satisfies the conditions of Theorems 1.3.2 and 1.3.3, and is a local operator.*

Proof. We note that
$$\text{Quad}(H) = W^{1,2}(\Omega)$$
and that
$$Q(f) = \int_\Omega \sum a_{ij} \frac{\partial f}{\partial x_i} \frac{\partial \bar{f}}{\partial x_j} dx$$
for all $f \in \text{Quad}(H)$. It follows from Lemma 1.2.9 and Theorem 1.2.10 that H satisfies condition (ii) of Theorem 1.3.2 and Condition (iv) of Theorem 1.3.3. To prove that H is local it is sufficient to observe that (1.3.8) holds trivially for all f in the dense subspace $C^\infty(\Omega) \cap W^{1,2}(\Omega)$ of $W^{1,2}(\Omega)$.

An advantage of the abstract formulation of the theory of this section is that it may also be applied to continuous time Markov chains and to the second order difference operators associated with random walks. To illustrate this let N be a countable set and let $A_{m,n}$ be a real symmetric matrix on $N \times N$ such that $m \neq n$ implies $A_{m,n} \leq 0$ and
$$0 \leq \sum_n A_{m,n} < \infty \quad (1.3.11)$$
for all $m \in N$. Let D be the space of functions of finite support on N, regarded

as a subspace of $l^2(N)$, where N is given the counting measure. Finally define the quadratic form Q on D by

$$Q(f) = \sum_{m,n} A_{m,n} f_n \bar{f}_m.$$

Theorem 1.3.10. *The quadratic form Q is closable on D, and its closure is associated with a self-adjoint extension H of the operator defined for all $f \in D$ by*

$$(Hf)_m = \sum_n A_{m,n} f_n. \qquad (1.3.12)$$

The operator H on $l^2(N)$ satisfies the conditions of Theorems 1.3.2 and 1.3.3.

Proof. The condition (1.3.11) implies that the operator H defined on D by (1.3.12) satisfies

$$Hf \in l^1(N) \subseteq l^2(N).$$

Moreover

$$\langle Hf, f \rangle = Q(f)$$

for all $f \in D$, so H is symmetric and Q is closable by Theorem 1.2.8.

Since $A_{m,n} \leq 0$ for all $m \neq n$, (1.3.11) implies that

$$Q(f) = \tfrac{1}{2} \sum_{m \neq n} |A_{m,n}| |f_m - f_n|^2 + \sum_m V_m |f_m|^2$$

where $V_m \geq 0$ for all $m \in N$. It is immediate from this representation that Q satisfies Condition (ii) of Theorem 1.3.2 and Condition (iii) of Theorem 1.3.3 on D, and hence also on $\text{Quad}(H)$.

We remark that the operator H above is certainly not local in the sense of (1.3.7). We next describe a variant of the above example, arising from graph theory, where each point is given its own positive weight.

Example 1.3.11. Let V denote the set of vertices of a graph and write $x \sim y$ if (x, y) is an unoriented edge. Assume that the number $n(x)$ of edges at each $v \in V$ is finite and define the Laplace operator by

$$\Delta f(x) = \frac{1}{n(x)} \sum_{y \sim x} \{f(y) - f(x)\}$$

for all f of finite support in V. If $n(x) = q$ for all $x \in V$ then

$$\Delta = q^{-1} A - 1$$

where the adjacency matrix A is defined by

$$A(x, y) = \begin{cases} 1 & \text{if } x \sim y \\ 0 & \text{otherwise.} \end{cases}$$

Symmetric Markov semigroups

If we take our Hilbert space to be

$$l^2(V, n) = \{f : \sum_x n(x)|f(x)|^2 < \infty\}$$

then $-\Delta$ is non-negative and symmetric with associated form

$$Q(f) = \tfrac{1}{2} \sum_{x \sim y} |f(x) - f(y)|^2.$$

It is evident as in Theorem 1.3.10 that the operator H associated with the quadratic form Q satisfies the conditions of Theorems 1.3.2 and 1.3.3.

If we let E denote the set of oriented edges $e = (x, y)$ of the graph, then we can define the operator $d : l^2(V, n) \to l^2(E)$ by

$$df(e) = f(y) - f(x).$$

It is easy to see that

$$-\Delta = d^*d$$

and that

$$0 \leqslant Q(f) \leqslant \sum_{x \sim y} (|f(x)|^2 + |f(y)|^2)$$
$$= 2 \sum_x n(x)|f(x)|^2$$
$$= 2 \|f\|^2.$$

Therefore $\|d\| \leqslant 2^{\frac{1}{2}}$ and $\text{Sp}(-\Delta) \subseteq [0, 2]$.

In the particular case $V = \mathbb{Z}^N$ with $x \sim y$ if and only if $|x - y| = 1$, it is straightforward to prove that $\text{Sp}(-\Delta) = [0, 2]$, but this is by no means always true.

Many of the properties of second order elliptic operators have simple discrete analogues for graphs. For example the Harnack inequality may be expressed as follows. If $f \geqslant 0$ and $\Delta f \leqslant 0$ on V then

$$n(y)^{-1} f(x) \leqslant f(y) \leqslant n(x) f(x)$$

whenever $x \sim y$. We omit the elementary proof.

1.4 Symmetric Markov semigroups

In this section we use the Beurling–Deny conditions to derive properties of the semigroup e^{-Ht} for a second order elliptic operator H, which have no analogue for higher order elliptic operators. These properties are related to the fact that there is an underlying probabilistic interpretation related to a Markov process on Ω, which we shall not investigate explicitly. Analytically the section allows us to use the scale of L^p spaces in a way similar to the way that the Sobolev spaces are used for higher order elliptic operators. The advantage of this is that one is able to control operators which have measurable coefficients rather easily.

Let Ω be a set with a countably generated σ-field and a σ-finite measure dx. If $H \geqslant 0$ is a self-adjoint operator on $L^2(\Omega)$ satisfying the conditions of Theorems 1.3.2 and 1.3.3, then H is called a Dirichlet form and e^{-Ht} is called a symmetric Markov semigroup.

Theorem 1.4.1. *If e^{-Ht} is a symmetric Markov semigroup on $L^2(\Omega)$ then $L^1 \cap L^\infty$ is invariant under e^{-Ht}, and e^{-Ht} may be extended from $L^1 \cap L^\infty$ to a positive one-parameter contraction semigroup $T_p(t)$ on L^p for all $1 \leqslant p \leqslant \infty$. These semigroups are strongly continuous if $1 \leqslant p < \infty$, and are consistent in the sense that*

$$T_p(t)f = T_q(t)f$$

if $f \in L^p \cap L^q$. They are self-adjoint in the sense that

$$T_p(t)^* = T_q(t)$$

if $1 \leqslant p < \infty$ and $p^{-1} + q^{-1} = 1$.

Proof. If $f \in L^1 \cap L^\infty$, which is a subset of $L^2 \cap L^\infty$, then

$$\|e^{-Ht}\|_\infty \leqslant \|f\|_\infty$$

for all $t \geqslant 0$ by hypothesis. If also $g \in L^1 \cap L^\infty$ then

$$|\langle e^{-Ht}f, g\rangle| = |\langle f, e^{-Ht}g\rangle|$$
$$\leqslant \|f\|_1 \|e^{-Ht}g\|_\infty$$
$$\leqslant \|f\|_1 \|g\|_\infty.$$

By choosing g appropriately we deduce that

$$\|e^{-Ht}f\|_1 \leqslant \|f\|_1.$$

The proof of the first statement of the theorem is completed by applying the Riesz–Thorin interpolation theorem.

We next prove that e^{-Ht} is strongly continuous on $L^1(\Omega)$. If $f \geqslant 0$ is a bounded function which vanishes outside a set E of finite measure then

$$\lim_{t \to 0} \|\chi_E e^{-Ht}f\|_1 = \lim_{t \to 0} \langle e^{-Ht}f, \chi_E\rangle$$
$$= \langle f, \chi_E\rangle = \|f\|_1$$

by the L^2 continuity of $e^{-Ht}f$. But $\|e^{-Ht}f\|_1 \leqslant \|f\|_1$ so

$$\lim_{t \to 0} \|\chi_{\Omega \setminus E} e^{-Ht}f\|_1 = 0.$$

This implies that

$$\lim_{t\to 0} \|e^{-Ht}f - f\|_1 \leq \lim_{t\to 0} \|\chi_E(e^{-Ht}f - f)\|_1$$
$$= \lim_{t\to 0} \langle |e^{-Ht}f - f|, \chi_E\rangle$$
$$\leq \lim_{t\to 0} \|e^{-Ht}f - f\|_2 |E|^{\frac{1}{2}} = 0$$

for such f, which are dense in L^1_+.

The strong continuity of e^{-Ht} on L^1 and L^2 implies the same for L^p with $1 < p < 2$ by interpolation. The same now follows for $2 < p < \infty$ by using the fact that L^p is reflexive for $1 < p < \infty$, and Theorem 1.34 of Davies (1980).

Although the semigroups $T_p(t)$ on L^p constructed in Theorem 1.4.1 are strongly continuous for $1 \leq p < \infty$ and have distinct generators H_p, these generators are consistent on the intersections of their domains. Although we shall often use the symbol H to stand for all these generators collectively we warn the reader that the spectrum of H_p may be entirely different for two different values of p.

Since the semigroup $T_\infty(t)$ on L^∞ is not generally strongly continuous we define its generator by

$$(\lambda + H_\infty)^{-1} = \{(\lambda + H_1)^{-1}\}^*$$

for all $\lambda > 0$, but comment that H_∞ does not normally have a domain dense in L^∞.

Theorem 1.4.2. *If e^{-Ht} is a symmetric Markov semigroup on $L^2(\Omega)$, $T_p(t)$ is a bounded holomorphic semigroup on $L^p(\Omega)$ for $1 < p < \infty$ with angle*

$$\theta_p \geq (\pi/2)(1 - |2/p - 1|). \tag{1.4.1}$$

Proof. Let $r > 0$, $-\pi/2 < \theta < \pi/2$, $f \in L^1 \cap L^2$ and $g \in L^2 \cap L^\infty$. Then consider the operator A_z defined on the strip $\{z : 0 \leq \operatorname{Re} z \leq 1\}$ by

$$\langle A_z f, g\rangle = \langle e^{-Hh(z)}f, g\rangle$$

where $h(z) = re^{i\theta z}$. By the L^2 spectral theorem the LHS is bounded on the strip. Moreover

$$|\langle A_z f, g\rangle| \leq \begin{cases} \|f\|_1 \|g\|_\infty & \text{if } \operatorname{Re} z = 0 \\ \|f\|_2 \|g\|_2 & \text{if } \operatorname{Re} z = 1. \end{cases}$$

By the Stein interpolation theorem we deduce that

$$\|A_t f\|_{p(t)} \leq \|f\|_{p(t)}$$

for all $0 < t < 1$, where

$$p(t)^{-1} = 1 - t + t/2.$$

Equivalently if $1 < p < 2$ we obtain

$$\|e^{-H\zeta}f\|_p \leqslant \|f\|_p$$

provided

$$|\text{Arg}\,\zeta| \leqslant (\pi/2)(2 - 2/p).$$

This bound coincides with (1.4.1) for such p. If $2 < p < \infty$ the corresponding bound is obtained by duality.

The above theorem can sometimes be extended to include the case $p = 1$ and sometimes cannot; see Section 4.3 for an example where it cannot be so extended.

If e^{-Ht} is a self-adjoint, positivity-preserving semigroup on $L^2(\Omega)$ we say that a Borel set E is invariant if $f = f\chi_E$ implies $e^{-Ht}f = (e^{-Ht}f)\chi_E$ for all $t \geqslant 0$, modulo null sets. This notion is independent of p. We say that e^{-Ht} is irreducible if Ω and $\emptyset$ are the only invariant sets, modulo null sets. The following result is well known.

Proposition 1.4.3. *If there is a positive integral kernel $K(t, x, y)$ on $(0, \infty) \times \Omega \times \Omega$ such that*

$$e^{-Ht}f(x) = \int_\Omega K(t, x, y)f(y)\,dy$$

for all $0 \leqslant f \in L^2$ then e^{-Ht} is irreducible. If e^{-Ht} is an irreducible self-adjoint positivity-preserving semigroup and E is the bottom of the spectrum of H then

$$\mathscr{L} = \{f \in \text{Dom}(H) : Hf = Ef\}$$

has dimension zero or one; in the latter case $\mathscr{L} = \mathbb{R}f$ where f is strictly positive almost everywhere.

It is a fundamental fact that irreducibility holds in great generality for a second order elliptic operator on a region (open connected set) in $\mathbb{R}^N$; see Theorem 3.3.5. The corresponding result for Schrödinger operators may be proved using the Trotter product formula or the Feynman–Kac formula.

In our above definition of symmetric Markov semigroups we did not assume that $T_t 1 = 1$. This conservation of probability condition is usually inappropriate for elliptic operators satisfying Dirichlet boundary conditions, but is sometimes valid. We have the following abstract result.

Theorem 1.4.4. *Let $T_t = e^{-Ht}$ be a symmetric Markov semigroup with resolvent*

$$R_\lambda = (H + \lambda)^{-1}$$

defined on L^p for all $1 \leqslant p \leqslant \infty$ and all $\lambda > 0$. Then the following conditions are equivalent:

(i) $T_t 1 = 1$ for some $t > 0$

(ii) $T_t 1 = 1$ for all $t > 0$

(iii) $R_\lambda 1 = \lambda^{-1} 1$ for some $\lambda > 0$

(iv) $R_\lambda 1 = \lambda^{-1} 1$ for all $\lambda > 0$.

Proof. (i)$\Rightarrow$(ii) The semigroup property implies $T_{nt} 1 = 1$ for all integers $n \geqslant 1$. If $s \geqslant 0$ and we choose n so that $0 \leqslant s \leqslant nt$ then

$$1 = T_{nt} 1 = T_{nt-s} T_s 1$$
$$\leqslant T_s \leqslant 1$$

so $T_s = 1$.

(ii)$\Rightarrow$(iv) This is an immediate consequence of the formula

$$R_\lambda 1 = \int_0^\infty e^{-\lambda t} T_t 1 \, dt$$

interpreted in the weak* sense on L^∞.

(iii)$\Rightarrow$(i) We have

$$0 = \int_0^\infty e^{-\lambda t} (1 - T_t 1) \, dt$$

where the integrand is non-negative, so $T_t 1 = 1$ for almost every $t > 0$. The proof is completed by observing that (iv)$\Rightarrow$(iii) is trivial.

1.5 An inequality of Hardy

In this section we prove an extension to several dimensions of an inequality due to Hardy which is closely related to the uncertainty principle. This inequality will enable us to obtain lower bounds on the spectrum of elliptic operators satisfying Dirichlet boundary conditions. It will also be a crucial ingredient in our analysis of the boundary behaviour of heat kernels in Chapter 4.

Lemma 1.5.1. *If* $f:[a,b] \to \mathbb{C}$ *is continuously differentiable with* $f(a) = f(b) = 0$ *then*

$$\int_a^b \frac{|f(x)|^2}{4d(x)^2} \, dx \leqslant \int_a^b |f'(x)|^2 \, dx$$

where

$$d(x) = \min\{|x-a|, |x-b|\}.$$

Proof. It is sufficient to prove that

$$\int_a^c \frac{|f(x)|^2}{4(x-a)^2} dx \leq \int_a^c |f'(x)|^2 dx$$

where $2c = a + b$, and a similar inequality for the other half-interval. It is sufficient to deal with the case where $a = 0$, and where f is real. We then have

$$\int_0^c (f')^2 dx = \int_0^c \{x^{\frac{1}{2}}(x^{-\frac{1}{2}}f)' + (1/2x)f\}^2 dx$$

$$\geq \int_0^c \{x^{-\frac{1}{2}}f(x^{-\frac{1}{2}}f)' + f^2/4x^2\} dx$$

$$= [\tfrac{1}{2}(x^{-\frac{1}{2}}f)^2]_0^c + \int_0^c (f^2/4x^2) dx$$

$$= f(c)^2/2c + \int_0^c (f^2/4x^2) dx$$

$$\geq \int_0^c (f^2/4x^2) dx$$

as required.

If Ω is a region in $\mathbb{R}^N$ and $\|u\| = 1$ we define

$$d_u(x) = \min\{|t| : x + tu \notin \Omega\}$$

so that $0 \leq d_u(x) \leq +\infty$ and $d_u(x) = 0$ if and only if $x \notin \Omega$. We also put

$$d_i(x) = d_{e(i)}(x)$$

where $\{e(i)\}_{i=1}^N$ is some orthonormal basis of $\mathbb{R}^N$.

Lemma 1.5.2. *If* $f \in C_c^\infty(\Omega)$ *then*

$$\sum_{i=1}^N \int_\Omega \frac{|f(x)|^2}{4d_i(x)^2} dx \leq \int_\Omega |\nabla f|^2 dx. \qquad (1.5.1)$$

Proof. We may assume that $\{e(i)\}$ is the standard basis and use Lemma 1.5.1 to obtain

$$\int_\Omega \frac{|f(x)|^2}{4d_i(x)^2} dx \leq \int_\Omega \left|\frac{\partial f}{\partial x_i}\right|^2 dx$$

from which the lemma follows by summation.

We now define the function m on $\mathbb{R}^N$ by

$$\frac{1}{m(x)^2} = \int_{\|u\|=1} \frac{dS(u)}{d_u(x)^2} \qquad (1.5.2)$$

where dS is the surface measure on the unit sphere of $\mathbb{R}^N$, normalised to have

An inequality of Hardy

unit total mass. Clearly $m(x) = 0$ if $x \notin \bar{\Omega}$ and $m(x) > 0$ if $x \in \Omega$. Moreover we always have

$$m(x) \geq d(x) \equiv \min\{|x-y| : y \notin \Omega\}. \tag{1.5.3}$$

Theorem 1.5.3. *If H is a strictly elliptic operator on $L^2(\Omega)$ with $a(x) \geq \lambda > 0$ for all $x \in \Omega$ then*

$$H \geq \lambda N/4m^2$$

in the sense of quadratic forms, where $m(x)$ is defined by (1.5.2).

Proof. By averaging (1.5.1) over all orthonormal bases using the group $O(n)$ we obtain

$$\int_{\|u\|=1} \int_\Omega \frac{N|f(x)|^2}{4d_u(x)^2} \,dx\, dS(u) \leq \int_\Omega |\nabla f|^2 \,dx$$

for all $f \in C_c^\infty$, or equivalently

$$\int_\Omega \frac{\lambda N|f(x)|^2}{4m(x)^2} \,dx \leq \int_\Omega \lambda|\nabla f|^2 \,dx$$

$$\leq Q(f) \tag{1.5.4}$$

where Q is the form of H. Since C_c^∞ is a form core of H, for every $f \in \text{Quad}(H)$ there exist $f_n \in C_c^\infty$ with $\|f_n - f\| \to 0$ and $Q(f_n) \to Q(f)$. The result now follows by the lower semicontinuity of the form associated with the LHS of (1.5.4), or Fatou's lemma.

We now say that $\Omega \subseteq \mathbb{R}^N$ is a regular domain if there exist constants c_1, c_2 with $c_1 > 0$ such that

$$H_0 \geq c_1/d^2 - c_2 \tag{1.5.5}$$

as a quadratic form inequality, where $H_0 = -\Delta$ with Dirichlet boundary conditions in the usual sense and $d(x)$ is given by (1.5.3).

We also say that Ω satisfies a uniform external ball condition if there exist $\beta > 0$ and $\alpha > 0$ such that for any $y \in \partial\Omega$ and $0 < s \leq \beta$ there exists a ball $B_{b,r}$ with centre b satisfying $|b - y| \leq s$, and radius r satisfying $r \geq 2\alpha s$, which does not meet Ω. This condition is implied by a uniform external cone condition and also holds if $\partial\Omega$ satisfies a uniform Lipschitz condition.

Theorem 1.5.4. *If $\Omega \subseteq \mathbb{R}^N$ satisfies a uniform external ball condition then it is a regular domain.*

Proof. We show that $d(x) \leq \beta$ implies

$$\frac{1}{m(x)^2} \geq \frac{\omega(\alpha)}{4d(x)^2} \tag{1.5.6}$$

where
$$\omega(\alpha) = \tfrac{1}{2}\int_0^{\sin^{-1}\alpha} \sin^{N-2}(t)\,dt \bigg/ \int_0^{\pi/2} \sin^{N-2}(t)\,dt.$$

This yields the bound
$$\frac{1}{m^2} \geq \frac{\omega(\alpha)}{4}\left(\frac{1}{d^2} - \frac{1}{\beta^2}\right)$$
after which we can apply Theorem 1.5.3.

Given $x\in\Omega$ with $d(x) \leq \beta$ let $y\in\partial\Omega$ satisfy $|x-y| = d(x)$ and consider the ball $B_{b,r}$ mentioned above where $s = d(x)$. The ball has a radius of at least $2\alpha s$ and the distance of its centre from x is at most $2s$. Therefore it subtends a (normalised) solid angle of at least $\omega(\alpha)$ at x. Every ray through x within this solid angle meets $\partial\Omega$ at a distance of at most $2d(x)$ from x. The definition (1.5.2) now leads to the bound (1.5.6).

We now define the inradius of a region Ω in $\mathbb{R}^N$ by
$$\text{Inr}(\Omega) = \sup\{d(x): x\in\Omega\}.$$

We also say that Ω is a strongly regular domain if we may put $c_2 = 0$ in (1.5.5). This corresponds to putting $\beta = \infty$ in the uniform external ball condition.

Theorem 1.5.5. *If $\Omega \subseteq \mathbb{R}^N$ has finite inradius and satisfies a uniform external ball condition then it is a strongly regular domain with constant c_1 determined geometrically by*
$$c_1 = c_1(N, \alpha, \beta, \text{Inr}(\Omega)).$$

Proof. We follow the same proof as before except that we must now put
$$s = \min\{d(x), \beta\}.$$
It follows that for any $x\in\Omega$ the ball $B_{b,r}$ subtends an angle of at least
$$\min\left\{\omega(\alpha), \omega\left(\frac{2\alpha\beta}{\beta + \text{Inr}(\Omega)}\right)\right\}$$
at x. The remainder of the proof is as before.

Theorem 1.5.6. *If Ω is a regular domain then*
$$W_0^{1,2}(\Omega) = W^{1,2}(\Omega) \cap \left\{f: \int_\Omega \frac{|f|^2}{d^2}\,dx < \infty\right\}.$$

Proof. It only remains to show that if $f\in W^{1,2}(\Omega)$ and the integral on the RHS is finite then $f\in W_0^{1,2}(\Omega)$. If ϕ is a non-negative C^∞ function on $\mathbb{R}^N$ such that $\phi(x) = 1$ if $|x| \leq 1$ and $\phi(x) = 0$ if $|x| \geq 2$, then

$$f(x) = \lim_{n \to \infty} f(x)\phi(x/n)$$

pointwise and in the $W^{1,2}$ norm. Therefore it is sufficient to prove the theorem when f has bounded support in $\mathbb{R}^N$. We now put

$$f_n(x) = f(x)F(n\tilde{d}(x))$$

where F is a non-decreasing C^∞ function such that $F(s) = 0$ if $s \leq \frac{1}{2}$ and $F(s) = 1$ if $s \geq 1$, and where $\tilde{d}$ is a C^∞ function comparable to d in the sense of Section 1.1.12. It is evident that f_n has compact support in Ω and that $\|f_n - f\|_2 \to 0$. We also have

$$\|\nabla(f_n - f)\|_2 = \|(F(n\tilde{d}) - 1)\nabla f + nF'(n\tilde{d})f\nabla \tilde{d}\|_2$$
$$\leq \|(F(n\tilde{d}) - 1)\nabla f\|_2 + \|n\tilde{d}F'(n\tilde{d})(f/\tilde{d})\nabla \tilde{d}\|_2.$$

The first norm converges to zero by the dominated convergence theorem. The second is dominated by the square root of

$$c \int_{E(n)} (|f|^2/\tilde{d}^2) \, dx \qquad (1.5.7)$$

where

$$c = \sup\{|sF'(s)| : 0 \leq s < \infty\}$$

and

$$E(n) = \{x : 1/2n \leq \tilde{d}(x) \leq 1/n\}.$$

The finiteness of $\int_\Omega (|f|^2/\tilde{d}^2) \, dx$ implies that (1.5.7) converges to zero as $n \to \infty$.

In spite of the completeness of the above description of $W_0^{1,2}(\Omega)$ the following sufficient condition for a function to lie in $W_0^{1,2}(\Omega)$ is frequently useful.

Theorem 1.5.7. *If Ω is any domain in $\mathbb{R}^N$ and $f \in W^{1,2}(\Omega)$ and f vanishes on the boundary in the sense that*

$$\lim_{x \to a} f(x) = 0$$

for all $a \in \partial\Omega$, where we assume $x \in \Omega$ when taking the limit, then $f \in W_0^{1,2}(\Omega)$.

Proof. Using the decomposition $f = f_+ - f_-$, we see that it is sufficient to treat the case where $f \geq 0$.

Now

$$f(x) = \lim_{n \to \infty} f(x)\phi(x/n)$$

where ϕ is any non-negative C^∞ function with $\phi(x) = 1$ if $|x| \leq 1$ and $\phi(x) = 0$ if $|x| \geq 2$. This limit is a limit in the $W^{1,2}(\Omega)$ norm, so it is sufficient to prove the theorem in the case where $f \geq 0$ has bounded support in $\mathbb{R}^N$. If we now put

$$f_n = (f - 1/n)_+$$

we see that f_n has compact support in Ω and that $f_n \to f$ in the norm of $W^{1,2}(\Omega)$. Therefore $f \in W_0^{1,2}(\Omega)$.

We may also use the above theorems to prove spectral properties for elliptic operators. Let C_N denote the bottom of the spectrum of $-\Delta$ in the unit ball of $\mathbb{R}^N$ subject to Dirichlet boundary conditions; thus $C_1 = \pi^2/4$ and $C_2 = 5.78316\ldots$.

Theorem 1.5.8. *Let H be a uniformly elliptic operator satisfying Dirichlet boundary conditions on the strongly regular domain $\Omega \subseteq \mathbb{R}^N$. Then the bottom $E_0(H)$ of the spectrum of H satisfies*

$$\frac{\lambda c_1}{\operatorname{Inr}(\Omega)^2} \leq E_0(H) \leq \frac{\mu C_N}{\operatorname{Inr}(\Omega)^2}$$

where c_1 is given by (1.5.5) and $c_2 = 0$. In particular $E_0(H) > 0$ if and only if the function d on Ω is bounded.

Proof. If $K = -\Delta$ on Ω subject to Dirichlet boundary conditions then the uniform ellipticity implies

$$\lambda K \leq H \leq \mu K$$

and hence

$$\lambda E_0(K) \leq E_0(H) \leq \mu E_0(K).$$

If $r < \operatorname{Inr}(\Omega)$ then there exists a ball of radius r contained in Ω. If L is its Laplacian then

$$E_0(K) \leq E_0(L) = C_N r^{-2}.$$

Combining these inequalities yields the upper bound of the theorem. Secondly the strong regularity states that

$$K \geq \frac{c_1}{d^2} \geq \frac{c_1}{\operatorname{Inr}(\Omega)^2}$$

and this yields the lower bound.

It is significant that the regularity condition in Theorem 1.5.8 cannot be removed. In the following example we put $H = -\Delta$ and $N \geq 3$ for simplicity, but comment that a similar argument works for $N = 2$.

Example 1.5.9. Let $N \geq 3$ and $0 < \varepsilon < \tfrac{1}{8}$ and put

$$\Omega_\varepsilon = \{x \in \mathbb{R}^N : \operatorname{dist}(x, \mathbb{Z}^N) > \varepsilon\}.$$

An inequality of Hardy

Let E_ε be the bottom of the spectrum of $H = -\Delta$ in $L^2(\Omega_\varepsilon)$ subject to Dirichlet boundary conditions. We shall show that $E_\varepsilon \to 0$ as $\varepsilon \to 0$ even though

$$\lim_{\varepsilon \to 0} \mathrm{Inr}\,(\Omega_\varepsilon) = (N/2)^{\frac{1}{2}} > 0.$$

Although each Ω_ε is strongly regular the constant c_1 in Theorem 1.5.8 depends upon ε and converges to zero as $\varepsilon \to 0$. The limiting region Ω_0 has $E_0 = 0$ although $\mathrm{Inr}\,(\Omega_0) > 0$.

To prove the above statements let f be an increasing C^∞ function on $[0, \infty)$ such that $f(s) = 0$ if $s \leqslant 1$ and $f(s) = 1$ if $s \geqslant 2$. Let $g \in C_c^\infty(\mathbb{R}^N)$, let d_ε be the distance function for Ω_ε, and put

$$g_\varepsilon(x) = g(x) f(d_\varepsilon(x)/2\varepsilon).$$

Then $g_\varepsilon \in C_c^\infty(\Omega_\varepsilon)$ and

$$\lim_{\varepsilon \to 0} \int |g_\varepsilon|^2 = \int |g|^2 > 0$$

assuming that g does not vanish identically. Also

$$\int |\nabla g_\varepsilon|^2 = \int |f \nabla g + g \nabla f|^2$$

$$\leqslant (1+\varepsilon^{\frac{1}{2}}) \int |f \nabla g|^2 + (1+\varepsilon^{-\frac{1}{2}}) \int |g \nabla f|^2$$

$$\leqslant (1+\varepsilon^{\frac{1}{2}}) \int |\nabla g|^2 + \frac{1+\varepsilon^{-\frac{1}{2}}}{4\varepsilon^2} \int |g|^2 f'(d_\varepsilon/2\varepsilon)^2.$$

Since only a finite number of points of $\mathbb{Z}^N$ lie within the support of g, we see, using the fact that $N \geqslant 3$, that

$$\limsup_{\varepsilon \to 0} \int |\nabla g_\varepsilon|^2 \leqslant \int |\nabla g|^2.$$

Therefore

$$\limsup_{\varepsilon \to 0} E_\varepsilon \leqslant \limsup_{\varepsilon \to 0} \int |\nabla g_\varepsilon|^2 \Big/ \int |g_\varepsilon|^2$$

$$\leqslant \int |\nabla g|^2 \Big/ \int |g|^2$$

for all $g \in C_c^\infty(\mathbb{R}^N)$. But the RHS can be arbitrarily small for suitable g, so $E_\varepsilon \to 0$ as $\varepsilon \to 0$.

The most powerful way of using Theorem 1.5.3 to obtain a lower bound on the bottom of the spectrum is simply to observe that

$$E_0(H) \geqslant \lambda N/4R^2$$

where
$$R = \text{ess sup}\{m(x) : x \in \Omega\}.$$
This requires no regularity of Ω. As a typical example we mention the dense, infinitely connected region $\Omega \subseteq \mathbb{R}^2$ defined by
$$\Omega = \{x : x_i \notin \mathbb{Z} \text{ for } i = 1, 2\} \cup \{x : |x - y| < 1/8 \text{ for some } y \in \mathbb{Z}^2\}$$
for which the bottom of the spectrum is strictly positive.

A more interesting result for regions in $\mathbb{R}^2$ which need not have regular boundaries is given next. It is not valid as it stands in higher dimensions.

Theorem 1.5.10. *Let $H = -\Delta$ with Dirichlet boundary conditions on a simply connected region Ω in $\mathbb{R}^2$. Then*
$$H \geqslant 1/16d^2.$$

Proof. By the Riemann mapping theorem there exists an analytic function f mapping the half-plane $U = \{x + iy : x > 0\}$ one–one onto Ω. By Koebe's one-quarter theorem
$$d(f(z)) \geqslant (x/2)|f'(z)|$$
for all $z \in U$. If $\phi \in C_c^\infty(\Omega)$ then by Lemma 1.5.1
$$\int_\Omega |\nabla \phi(z)|^2 \, dx \, dy = \int_U |\nabla \phi(f(z))|^2 \, dx \, dy$$
$$\geqslant \int_U \{|\phi(f(z))|^2/4x^2\} \, dx \, dy$$
$$\geqslant \int_\Omega \{|\phi(z)|^2/16d^2\} \, dx \, dy$$

The following theorem is less powerful than Theorem 1.5.3, but possibly easier to interpret.

Theorem 1.5.11. *Suppose that H is strictly elliptic on Ω and that there exist $r > 0$ and $0 < \gamma < 1$ such that the ball $B_{x,r}$ with centre x and radius r satisfies*
$$|B_{x,r} \cap \Omega| \leqslant \gamma |B_r|$$
for $x \in \mathbb{R}^N$. Then the bottom $E_0(H)$ of the spectrum of H satisfies
$$E_0(H) \geqslant c(\gamma, N)\lambda/r^2$$
where $c(\gamma, N) > 0$ and $\lambda > 0$ is the constant of strict ellipticity.

Proof. Given $x \in \mathbb{R}^N$ let
$$S_x = \{u : \|u\| = 1 \text{ and } d_u(x) < \mu\}$$

where
$$0 < \mu = r\left(\frac{2\gamma}{1+\gamma}\right)^{1/N} < r.$$
Then
$$|B_{x,r} \cap \Omega| \geq (1 - |S_x|) v_N \mu^N$$
where v_N is the volume of the unit ball in $\mathbb{R}^N$, and the area of the unit sphere is normalised to unity. Applying the hypothesis of the theorem leads to
$$1 - |S_x| \leq \gamma r^N/\mu^N = (1+\gamma)/2$$
and hence
$$|S_x| \geq (1-\gamma)/2.$$
Theorem 1.5.3 states that
$$H \geq \lambda N/4m^2$$
where
$$m(x)^{-2} = \int_{\|u\|=1} dS(u)/d_u(x)^2$$
$$\geq \mu^{-2}|S_x|$$
$$= (1-\gamma)/2\mu^2.$$
Combining these yields
$$H \geq \left(\frac{1+\gamma}{2\gamma}\right)^{2/N} \frac{(1-\gamma)\lambda N}{r^2}$$
as required for the theorem.

Although it is possible to adapt the proof of Theorem 1.5.3 to elliptic operators H which are not strictly elliptic, there is another approach which we now describe. The value of this method depends upon a suitable choice of $\phi > 0$.

Theorem 1.5.12. *Suppose that H is elliptic on $L^2(\Omega)$ and that there is a positive continuous function ϕ in $W^{1,2}_{\text{loc}}(\Omega)$ and a potential V in $L^1_{\text{loc}}(\Omega)$ such that*
$$H\phi \geq V\phi$$
in the weak sense that
$$\int_\Omega \sum_{i,j} a_{ij} \frac{\partial \phi}{\partial x_i} \frac{\partial u}{\partial x_j} dx \geq \int V\phi u \, dx \quad (1.5.8)$$
for all $0 \leq u \in C_c^\infty(\Omega)$. Then the quadratic form inequality
$$H \geq V$$
is valid on $C_c^\infty(\Omega)$.

Proof. If $f \in C_c^\infty(\Omega)$ then we may put $f = \phi g$ where $g \in W_c^{1,2}(\Omega)$ and obtain

$$\int_\Omega \sum a_{ij} \frac{\partial f}{\partial x_i} \frac{\partial \bar{f}}{\partial x_j} dx = \int_\Omega \sum a_{ij} \left(\frac{\partial \phi}{\partial x_i} g + \phi \frac{\partial g}{\partial x_i} \right) \left(\frac{\partial \phi}{\partial x_j} \bar{g} + \phi \frac{\partial \bar{g}}{\partial x_j} \right) dx$$

$$\geq \int_\Omega \sum a_{ij} \left(|g|^2 \frac{\partial \phi}{\partial x_i} \frac{\partial \phi}{\partial x_j} + \phi g \frac{\partial \phi}{\partial x_i} \frac{\partial \bar{g}}{\partial x_j} + \phi \bar{g} \frac{\partial g}{\partial x_i} \frac{\partial \phi}{\partial x_j} \right) dx$$

$$= \int_\Omega \sum a_{ij} \frac{\partial \phi}{\partial x_i} \frac{\partial}{\partial x_j} (\phi |g|^2) dx$$

$$\geq \int_\Omega V \phi^2 |g|^2 dx = \int V |f|^2 dx$$

since $\phi |g|^2$ may be approximated in a suitable sense by a sequence $0 \leq u_n \in C_c^\infty$, by using a mollifier.

We call a function ϕ in $W_{loc}^{1,2}(\Omega)$ $(H-V)$-superharmonic when

$$(H-V)\phi \geq 0$$

in the sense of (1.5.8). We also say that ϕ is subharmonic if $-\phi$ is superharmonic and harmonic if $\pm \phi$ are both superharmonic. We shall not pursue the potential theory of such operators, but note the following result.

Proposition 1.5.13. *The class of $(H-V)$-superharmonic functions on Ω is a cone invariant under* min.

The following is a typical application of Theorem 1.5.12. See also Theorem 1.6.6 for a related result.

Theorem 1.5.14. *Let H be an elliptic operator on $L^2(\mathbb{R}^N)$ whose coefficients satisfy*

$$a(x) \geq \lambda(1 + x^2) \quad (1.5.9)$$

for some $\lambda > 0$. Then

$$\mathrm{Sp}(H) \subseteq [\tfrac{1}{4}\lambda N^2, \infty).$$

Proof. By monotonicity of the spectrum it is sufficient to treat the case where one has equality in (1.5.9). Putting

$$\phi(x) = (1 + x^2)^{-\frac{1}{4}N}$$

we have

$$H\phi = \tfrac{1}{4} N \nabla \cdot \{\lambda (1 + x^2)^{-\frac{1}{4}N} 2x\}$$
$$= \tfrac{1}{4} \lambda N \{(1 + x^2)^{-\frac{1}{4}N} 2N - \tfrac{1}{4} N(1 + x^2)^{-\frac{1}{4}N - 1} 4x^2\}$$
$$= \tfrac{1}{4} \lambda N^2 \phi \{2 - x^2/(1 + x^2)\}.$$

Theorem 1.5.12 now implies that

$$H \geqslant \tfrac{1}{4}\lambda N^2 \frac{2+x^2}{1+x^2} \geqslant \tfrac{1}{4}\lambda N^2.$$

1.6 Compactness and spectrum

We start with a general result concerning compactness and interpolation. We assume that Ω is a set with a countably generated σ-field and a σ-finite measure dx.

Theorem 1.6.1. *Suppose that $1 \leqslant p_0 < p < p_1 \leqslant \infty$ and that the linear operator $A: L^{p_0} \cap L^{p_1} \to L^{p_0} \cap L^{p_1}$ can be extended to a bounded linear operator from L^{p_1} to L^{p_1} and to a compact linear operator from L^{p_0} to L^{p_0}. Then A can be extended to a compact linear operator from L^p to L^p.*

Proof. If $\{E_1, \ldots, E_n\}$ is a sequence of disjoint subsets of finite positive measure, then there is a projection P defined on L^q for all $1 \leqslant q \leqslant \infty$ by

$$Pf = \sum_{r=1}^{n} |E_r|^{-1} \int_{E_r} f(x)\,dx.$$

This projection is of finite rank and is a contraction on L^q for all $1 \leqslant q \leqslant \infty$. The hypothesis on Ω ensures that there is a sequence P_n of such projections such that P_n converges strongly to the identity operator on L^q for all $1 \leqslant q < \infty$.

Since A is compact on L^{p_0} we have

$$\lim_{n \to \infty} \|A - P_n A\|_{p_0, p_0} = 0$$

in an obvious notation. But

$$\|A - P_n A\|_{p_1, p_1} \leqslant 2\|A\|_{p_1, p_1}$$

so the Riesz–Thorin interpolation theorem implies that

$$\lim_{n \to \infty} \|A - P_n A\|_{p, p} = 0$$

for all $p_0 < p < p_1$. But $P_n A$ is of finite rank so A is compact on L^p.

Corollary 1.6.2. *Under the above circumstances the spectrum of A is the same for all p such that $p_0 \leqslant p < p_1$, and the spectral projections corresponding to non-zero eigenvalues are independent of p.*

Proof. Let A_p denote the extension of A from $L^{p_0} \cap L^{p_1}$ to L^p, and define A_{p_0} similarly. The set

$$S = \mathrm{Sp}(A_{p_0}) \cup \mathrm{Sp}(A_p) \subseteq \mathbb{C}$$

is countable and closed with 0 as its only possible limit point. If $f \in L^{p_0} \cap L^p$ then

$$(A_p - z)^{-1} f = (A_{p_0} - z)^{-1} f$$

for large $|z|$ by the power series expansion, and then for all $z \notin S$ by analytic continuation. If $0 \neq s \in S$ and γ is a small enough circle with centre s then the spectral projection for s is given for such f by

$$P_p f = (1/2\pi i) \int_\gamma (z - A_p)^{-1} f \, dz$$

$$= (1/2\pi i) \int_\gamma (z - A_{p_0})^{-1} f \, dz = P_{p_0} f.$$

Since $L^{p_0} \cap L^p$ is norm dense in both L^{p_0} and L^p, we see that P_p and P_{p_0} have the same finite-dimensional range. Hence $s \in \text{Sp}(A_p)$ if and only if $s \in \text{Sp}(A_{p_0})$. The number $s = 0$ always lies in $\text{Sp}(A_p)$ and $\text{Sp}(A_{p_0})$ unless Ω is finite, in which case the theorem is trivial.

We now apply the above theorems.

Theorem 1.6.3. *Let e^{-Ht} be a symmetric Markov semigroup for which e^{-Ht} is compact on $L^2(\Omega)$ for all $t > 0$. Then e^{-Ht} is compact on $L^p(\Omega)$ for all $t > 0$ and $1 < p < \infty$. The spectrum of H_p is independent of p for $1 < p < \infty$, and every L^2 eigenfunction of H_2 lies in L^p for all $1 < p < \infty$.*

Proof. This follows from the above theorems by putting $p_0 = 2$ and $p_1 = 1$ or ∞.

We shall give an example in Section 4.3 of an elliptic operator for which the above theorem applies, but for which $e^{-H_1 t}$ is not compact for any $t > 0$ and for which $\text{Sp}(H_1)$ is completely different from $\text{Sp}(H_2)$. However in other cases the following theorem is applicable.

Theorem 1.6.4. *Let e^{-Ht} be a symmetric Markov semigroup such that e^{-Ht} is compact on $L^1(\Omega)$ for all $t > 0$. Then e^{-Ht} is compact on $L^p(\Omega)$ for all $t > 0$ and $1 \leq p \leq \infty$. The spectrum of H_p is independent of p for $1 \leq p \leq \infty$ and every L^2 eigenfunction of H_2 lies in L^p for all $1 \leq p \leq \infty$.*

Proof. One puts $p_0 = 1$, $p_1 = \infty$ and $A = e^{-Ht}$ in Theorem 1.6.1 and Corollary 1.6.2. Note that H_∞ does not normally have a dense domain, but we can still use the identity

$$(\lambda + H_\infty)^{-1} = \{(\lambda + H_1)^{-1}\}^*$$

to analyse its spectral properties.

Criteria for the application of Theorem 1.6.4 to elliptic operators will be given in Section 2.1, and we concentrate here on the application of Theorem 1.6.3. For the remainder of this section we suppose that Ω is a region in $\mathbb{R}^N$ and that H is an elliptic differential operator on $L^2(\Omega)$ satisfying Dirichlet boundary conditions. For analogous results concerning Neumann boundary conditions see Section 1.7.

The following technical lemma will be of great value.

Lemma 1.6.5. *If K is strictly elliptic on $L^2(\Omega)$, and the potential V in $L^1_{\text{loc}}(\Omega)$ is bounded below, and the potential W is bounded with bounded support in Ω, then*

$$(K + V + \gamma)^{-1} - (K + V + W + \gamma)^{-1}$$

is compact for large enough $\gamma > 0$.

Proof. We choose $\gamma > 0$ large enough so that $V + \gamma \geqslant 1$ and $V + W + \gamma \geqslant 1$. Since K is strictly elliptic there exists $\lambda > 0$ such that $K \geqslant \lambda H_0$ on $L^2(\mathbb{R}^N)$, where $H_0 = -\Delta$. This quadratic form inequality is interpreted by extending the quadratic from Q of K form $L^2(\Omega)$ to $L^2(\mathbb{R}^N)$ by putting

$$Q(f) = \begin{cases} \langle K^{\frac{1}{2}} f, K^{\frac{1}{2}} f \rangle & \text{if } f \in \text{Quad } K \\ +\infty & \text{otherwise} \end{cases}$$

so that in particular $Q(f) = +\infty$ if $f \notin L^2(\Omega)$. We interpret the pseudo-resolvent $(K + V + \gamma)^{-1}$ on $L^2(\mathbb{R}^N)$ as being zero on $L^2(\mathbb{R}^N \setminus \Omega)$.

We have

$$(K + V + \gamma)^{-1} - (K + V + W + \gamma)^{-1}$$
$$= (K + V + W + \gamma)^{-1} W (K + V + \gamma)^{-1}$$
$$= (K + V + W + \gamma)^{-1} AB (K + V + \gamma)^{-\frac{1}{2}}$$

where

$$A = W(\lambda H_0 + 1)^{-\frac{1}{2}}$$

and

$$B = (\lambda H_0 + 1)^{\frac{1}{2}} (K + V + \gamma)^{-\frac{1}{2}}.$$

Our proof will be completed by proving that A is compact and B is a contraction.

If we put

$$A_\varepsilon = W e^{-\varepsilon H_0} (\lambda H_0 + 1)^{-\frac{1}{2}}$$

where $\varepsilon > 0$, then A_ε has a kernel of the form

$$W(x) g_\varepsilon(x - y)$$

where $g_\varepsilon \in L^2(\mathbb{R}^N)$ by Section 1.1.9. Therefore A_ε is Hilbert–Schmidt. But

$\|A_\varepsilon - A\| \to 0$ as $\varepsilon \to 0$, so A is compact. Finally
$$0 \leq B^*B = (K+V+\gamma)^{-\frac{1}{2}}(\lambda H_0 + 1)(K+V+\gamma)^{-\frac{1}{2}} \leq 1$$
by virtue of the quadratic form inequality
$$0 \leq \lambda H_0 + 1 \leq K + V + \gamma$$
on $L^2(\mathbb{R}^N)$. Hence B is a contraction as claimed.

Theorem 1.6.6. *If the coefficients $a(x)$ of the elliptic operator H on $L^2(\mathbb{R}^N)$ satisfy*
$$\lim_{|x| \to \infty} \|a(x)\|/x^2 = 0 \tag{1.6.1}$$
then 0 lies in the essential spectrum of H. Conversely if $c > 0$ and
$$a(x) \geq cx^2 \tag{1.6.2}$$
for large $|x|$ then
$$\operatorname{Ess Sp}(H) \subseteq [cN^2/4, \infty).$$

Proof. Let ϕ be a non-negative function in $C_c^\infty(\mathbb{R}^N)$ with $0 \leq \phi \leq 1$ and $\operatorname{supp}(\phi)$ in the unit ball. If we put $\phi_n(x) = \phi(x/n)$ then $\|\phi_n\|_2^2 = cn^N$ so
$$Q(\phi_n)/\|\phi_n\|^2 = c^{-1} n^{-N} \int \sum a_{ij} (\partial \phi_n/\partial x_i)(\partial \phi_n/\partial x_j) \, dx$$
$$\leq c^{-1} n^{-N} a^2 n^{-2} \int_{|x| \leq n} \sum_{i,j} |a_{ij}(x)| \, dx$$
$$\leq c_1 n^{-2} \sup\{\|a(x)\| : x \leq n\}$$
which converges to 0 as $n \to \infty$ by (1.6.1). We deduce that $0 \in \operatorname{Sp}(H)$. If 0 were an eigenvalue of H with corresponding eigenfunction $\psi \in L^2(\mathbb{R}^N)$ then we see by Theorem 1.2.6 that
$$0 = Q(\psi) = \int \sum_{i,j} a_{ij} (\partial \psi/\partial x_i)(\partial \bar\psi/\partial x_j) \, dx.$$
This implies $\nabla \psi = 0$ almost everywhere, and hence $\psi = 0$. Since this implies 0 is not an eigenvalue, it must lie in the essential spectrum.

For the second part we assume (1.6.2) holds and let λ be a positive C^∞ function on $\mathbb{R}^N$ such that $\lambda(x) 1 \leq a(x)$ for all $x \in \mathbb{R}^N$ and $\lambda(x) = cx^2$ for large $|x|$. We then define the operator K on $L^2(\mathbb{R}^N)$ by the usual form procedure starting from the formula
$$Kf = -\operatorname{div}(\lambda \operatorname{grad} f)$$
for all $f \in C_c^\infty$. Now let ϕ be a positive C^∞ function on $\mathbb{R}^N$ such that $\phi(x) = r^{-N/2}$ for large $r = |x|$. For such x we have

Compactness and spectrum

$$K\phi = \text{div}\{cr^2(N/2)r^{-1-N/2}(x/r)\}$$
$$= (cN^2/4)\phi$$

in the sense of differential equations. Note that $\phi \notin L^2(\mathbb{R}^N)$. Hence

$$(K + V)\phi(x) = 0$$

for all $x \in \mathbb{R}^N$, where the C^∞ potential V on $\mathbb{R}^N$ satisfies $V(x) = -cN^2/4$ for large $|x|$. Thus $V = W - cN^2/4$ where W is C^∞ with compact support.

By Theorem 1.5.12 we have

$$\text{Sp}(K + V) \subseteq [0, \infty)$$

so

$$\text{Sp}(K + W) \subseteq [cN^2/4, \infty).$$

But Lemma 1.6.5 implies that

$$(K + \gamma)^{-1} - (K + W + \gamma)^{-1}$$

is compact for suitable $\gamma > 0$. It follows by Davies (1980) Theorem 3.15 that the two resolvents have the same essential spectrum. Hence

$$\text{Ess Sp}(K) = \text{Ess Sp}(K + W)$$
$$\subseteq [cN^2/4, \infty).$$

But $H \geq K$, so the bottom of the essential spectrum of H is not less than $cN^2/4$.

Corollary 1.6.7. *If H is an elliptic operator on $L^2(\mathbb{R}^N)$ and the lowest eigenvalue $\lambda(x)$ of the $N \times N$ matrix $a(x)$ satisfies*

$$\lim \lambda(x)/x^2 = +\infty$$

then H has a compact resolvent.

Proof. The constant c in Theorem 1.6.6 may be taken arbitrarily large, so the essential spectrum of H is empty.

We now study a different type of condition for an elliptic operator to have empty essential spectrum.

Theorem 1.6.8. *Let H be uniformly elliptic on $L^2(\Omega)$ and satisfy Dirichlet boundary conditions. If $\Omega \subseteq \mathbb{R}^N$ is a regular domain, then H^{-1} is compact on $L^2(\Omega)$ if and only if*

$$\lim_{|x| \to \infty} d(x) = 0 \tag{1.6.3}$$

where

$$d(x) = \min\{|x - y| : y \notin \Omega\}.$$

Also H^{-1} is compact on $L^2(\Omega)$ for all bounded regions Ω, regular or not.

Proof. If (1.6.3) does not hold there exists a sequence $x_n \in \mathbb{R}^N$ such that $|x| \to \infty$ and $d(x_n) \geq 2\delta > 0$ for all n. Thus there exists a sequence of disjoint balls $B_n \subseteq \Omega$ of equal radius $\delta > 0$. Let ϕ_n be the eigenfunctions for $-\Delta$ in $L^2(B_n)$ with Dirichlet boundary conditions, corresponding to the lowest eigenvalue E, which is independent of n, and normalised by $\|\phi_n\|_2 = 1$. Then

$$\langle \phi_m, \phi_n \rangle = \delta_{m,n}.$$

Moreover $\phi_m \in W_0^{1,2}(B_m) \subseteq W_0^{1,2}(\Omega) = \text{Quad}(H)$ and

$$\langle H^{\frac{1}{2}} \phi_m, H^{\frac{1}{2}} \phi_n \rangle = 0$$

for $m \neq n$, by the locality of H. If $H_0 = -\Delta$ with Dirichlet boundary conditions in $L^2(\Omega)$, then there exists $\mu < \infty$ such that

$$\langle H^{\frac{1}{2}} \phi_m, H^{\frac{1}{2}} \phi_m \rangle \leq \mu \langle H_0^{\frac{1}{2}} \phi_m, H_0^{\frac{1}{2}} \phi_m \rangle \leq \mu E$$

by the uniform ellipticity of H. By the minimax method of calculating the eigenvalues of H, we deduce that there is a point of the essential spectrum of H within the interval $[0, E\mu]$, so the operator H^{-1} cannot be compact.

Conversely suppose (1.6.3) does hold. There exists $\lambda > 0$ such that $H \geq \lambda H_0$, by the uniform ellipticity of H. Using (1.5.5) we deduce that

$$H \geq \tfrac{1}{2}\lambda(H_0 + (c_1/d^2) - c_2) \tag{1.6.4}$$

where $c_1 > 0$.

The remainder of the proof is similar to that of Theorem 1.6.6. Because $d(x) \to 0$ as $|x| \to \infty$ with $x \in \Omega$, given $c > 0$ we have

$$H_0 + (c_1/d^2) - c_2 = H_0 + V + W \tag{1.6.5}$$

where $V(x) \geq c$ for all $x \in \Omega$ and W is continuous with bounded support. By Lemma 1.6.5 the essential spectrum of (1.6.5) equals that of $(H_0 + V)$, and hence is contained in $[c, \infty)$. But $c > 0$ is arbitrary so $(H_0 + V + W)$ has no essential spectrum. Applying the minimax argument to (1.6.4) we finally conclude that H has no essential spectrum.

Example 1.6.9. We remark that Theorem 1.6.8 applies not only to bounded regions but also to certain unbounded regions of infinite volume such as

$$\{(x, y) \in \mathbb{R}^2 : |y| \leq (1 + x^2)^{-\alpha}\}$$

where $0 < \tfrac{1}{2}$. If we assume Neumann boundary conditions, then the situation is different, and is treated in Theorem 1.7.12.

1.7 Some Sobolev inequalities

In this section we present a few Sobolev inequalities which are of fundamental importance.

Some Sobolev inequalities

If Ω is a region in $\mathbb{R}^N$ we have defined $W^{1,p}(\Omega)$ for $1 \leq p \leq \infty$ to be

$$\{f \in L^p(\Omega): \nabla f \in L^p(\Omega)\}$$

where ∇f is calculated in the weak sense. This is a Banach space for the norm

$$\|\|f\|\| = \{\|f\|_p^p + \|\nabla f\|_p^p\}^{1/p}.$$

By Lemma 1.2.9 $C^\infty(\Omega) \cap W^{1,p}(\Omega)$ is dense in $W^{1,p}(\Omega)$ for this norm; see also Lemma 1.7.11 below. We have shown in Lemma 1.2.4 that $W_0^{1,p}(\Omega)$ is the closure of $C_c^\infty(\Omega)$ in $W^{1,p}(\Omega)$.

Theorem 1.7.1. *If $1 \leq p < N$ and $1/q = 1/p - 1/N$ then*

$$W_0^{1,p}(\Omega) \subseteq L^q(\Omega)$$

and $f \in W_0^{1,p}(\Omega)$ implies

$$\|f\|_q \leq c \|\nabla f\|_p.$$

Proof. We may as well assume that $\Omega = \mathbb{R}^N$ since $W_0^{1,p}(\Omega) \subseteq W_0^{1,p}(\mathbb{R}^N)$. If $p = 1$ and $f \in C_c^1(\mathbb{R}^N)$ then

$$|f(x)| \leq \int_{-\infty}^\infty |\partial f/\partial x_i|\, dx_i$$

$$= g_i(x_1, \ldots, \hat{x}_i, \ldots, x_N)$$

where $\hat{x}_i$ indicates independence of the variable x_i. Therefore

$$|f(x)|^{N/(N-1)} \leq \prod_{i=1}^N g_i(x)^{1/(N-1)}$$

and

$$\int |f|^{N/(N-1)}\, dx_1 \leq g_1^{1/(N-1)} \int \prod_{i=2}^N g_i^{1/(N-1)}\, dx_1$$

$$\leq g_1^{1/(N-1)} \prod_{i=2}^N \left(\int g_i\, dx_1\right)^{1/(N-1)}$$

$$= \prod_{i=1}^N h_i^{1/(N-1)}$$

where $h_1 = g_1$ and

$$h_i = \int g_i\, dx_1$$

for $i = 2, \ldots, n$, so that h_2 is independent of x_2. Thus

$$\int |f|^{N/(N-1)}\, dx_1\, dx_2 \leq h_2^{1/(N-1)} \int \prod_{i \neq 2} h_i^{1/(N-1)}\, dx_2$$

$$\leq h_2^{1/(N-1)} \prod_{i \neq 2} \left(\int h_i\, dx_2\right)^{1/(N-1)}.$$

Repeating the process inductively we end up with the bound

$$\int_{\mathbb{R}^N} |f|^{N/(N-1)}\,dx \leq \prod_{i=1}^{N}\left(\int_{\mathbb{R}^N}\left|\frac{\partial f}{\partial x_i}\right|dx\right)^{1/(N-1)}$$

which implies that

$$\|f\|_{N/(N-1)} \leq \left(\prod_{i=1}^{N}\int_{\mathbb{R}^N}\left|\frac{\partial f}{\partial x_i}\right|dx\right)^{1/N}$$

$$\leq \frac{1}{N}\sum_{i=1}^{N}\int_{\mathbb{R}^N}\left|\frac{\partial f}{\partial x_i}\right|dx$$

$$\leq \frac{1}{N^{\frac{1}{2}}}\int_{\mathbb{R}^N}|\nabla f|\,dx.$$

If $\gamma > 1$ then putting $f = g^\gamma$ where $g \in C_c^1$ we obtain

$$\left\{\int|g|^{\gamma N/(N-1)}\,dx\right\}^{(N-1)/N} \leq (1/N^{\frac{1}{2}})\int \gamma|g|^{\gamma-1}|\nabla g|\,dx$$

$$\leq (\gamma/N^{\frac{1}{2}})\left(\int|g|^{(\gamma-1)r}\,dx\right)^{1/r}\left(\int|\nabla g|^p\,dx\right)^{1/p}$$

(1.7.1)

if $1/r + 1/p = 1$. If we make the choice

$$\gamma = (Np - p)/(N - p)$$

then

$$\gamma N/(N-1) = (\gamma - 1)r = q$$

so (1.7.1) becomes

$$\|g\|_q^\gamma \leq (\gamma/N^{\frac{1}{2}})\|g\|_q^{\gamma-1}\|\nabla g\|_p$$

and

$$\|g\|_q \leq \frac{(N-1)p}{(N-p)N^{\frac{1}{2}}}\|\nabla g\|_p.$$

The theorem now follows from the fact that C_c^1 is dense in $W_0^{1,p}$.

We next consider the case where $p > N$. We define the functions h_s on $\mathbb{R}^N$ for $1 < s < \infty$ by

$$h_s(x) = |x|^{-N/s}$$

and note that h_s nearly lies in $L^s(\mathbb{R}^N)$. In fact

$$h_s \in L^{s-\varepsilon} + L^{s+\varepsilon}$$

for any $\varepsilon > 0$; this fact will motivate our calculations but will not actually be used.

Lemma 1.7.2. *If $f \in W_0^{1,1}(\Omega)$ then*
$$|f(x)| \leq c(h_{N/(N-1)} * |\nabla f|)(x)$$
almost everywhere, where $c = c(N)$.

Proof. As before it is sufficient to treat the case $\Omega = \mathbb{R}^N$. Since the above inequality is preserved under norm limits in $W_0^{1,1}$ we need only consider the case $f \in C_c^1$. We then have
$$f(x) = -\int_0^\infty \frac{\partial}{\partial r} f(x + r\omega) \, dr$$
for any $|\omega| = 1$ and hence
$$|f(x)| \leq S_N^{-1} \int_{|\omega|=1} \int_0^\infty |\nabla f(x + r\omega)| \, dr \, d\omega$$
where S_N is the area of the unit sphere in $\mathbb{R}^N$. Therefore
$$|f(x)| \leq S_N^{-1} \int \frac{|\nabla f(x + r\omega)|}{r^{N-1}} r^{N-1} \, dr \, d\omega$$
$$= S_N^{-1}(h_{N/(N-1)} * |\nabla f|)(x).$$

Lemma 1.7.3. *Let $\Omega \subseteq \mathbb{R}^N$ be a bounded convex set with volume $|\Omega|$ and diameter β. If $f \in W^{1,1}(\Omega)$ then*
$$|f(x) - f_\Omega| \leq \frac{\beta^N}{N|\Omega|} (h_{N/(N-1)} * |\nabla f|)(x)$$
almost everywhere, where
$$f_\Omega = \frac{1}{|\Omega|} \int_\Omega f(y) \, dy$$

Proof. By Lemma 1.2.9 or Lemma 1.7.11 below it is sufficient to treat the case where $f \in W^{1,1}(\Omega) \cap C^1(\Omega)$. If $x, y \in \Omega$ then
$$f(x) - f(y) = -\int_0^\rho \frac{\partial}{\partial r} f(x + r\omega) \, dr$$
where $\omega = (y - x)/|y - x|$ and $\rho = |x - y|$. Therefore
$$|f(x) - f_\Omega| = \left| \frac{1}{|\Omega|} \int_\Omega \int_0^\rho \frac{\partial}{\partial r} f(x + r\omega) \, dr \, dy \right|$$
$$\leq \frac{1}{|\Omega|} \int_0^\rho |\nabla f(x + r\omega)| \, dr \, dy$$
$$= \frac{1}{|\Omega|} \int_{|\omega|=1} \int_{r=0}^\rho |\nabla f(x + r\omega)| \rho^{N-1} \, dr \, d\rho \, d\omega$$

where $0 < \rho(\omega) < \beta$ for all $|\omega| = 1$. Therefore

$$|f(x) - f_\Omega| \leq \frac{\beta^N}{N|\Omega|} \int_{|\omega|=1} \int_{s=0}^{r(\omega)} |\nabla f(x + r\omega)| dr \, d\omega$$

$$= \frac{\beta^N}{N|\Omega|} \int_\Omega r^{-(N-1)} |\nabla f(x+u)| du$$

$$= \frac{\beta^N}{N|\Omega|} (h_{N/(N-1)} * |\nabla f|)(x)$$

if $u = r\omega$.

Theorem 1.7.4. *If $p > N$ and $f \in W_0^{1,p}(\Omega)$ then $f \in C_0(\bar{\Omega})$ and*

$$\|f\|_\infty \leq c \|\nabla f\|_p \beta^{1 - N/p} \qquad (1.7.2)$$

where β is the diameter of Ω and $c = c(N, p)$.

Proof. Lemma 1.7.2 states that

$$|f| \leq ch_{N/(N-1)} * |\nabla f|$$

which implies

$$\|f\|_\infty \leq c \|h_{N/(N-1)}\|_q \|\nabla f\|_p$$

where $1/p + 1/q = 1$. Also $p > N$ implies

$$\|h_{N/(N-1)}\|_q^q = \int_\Omega r^{-(N-1)q + N - 1} dr \, d\omega$$

$$\leq S_N \frac{\beta^{N - (N-1)q}}{N - (N-1)q}$$

$$= S_N \frac{1 - 1/p}{1 - N/p} \beta^{(1 - N/p)q}$$

which yields the bound (1.7.2). If $f_n \in C_c^1(\Omega)$ and f_n converges to f in $W_0^{1,p}(\Omega)$ then (1.7.2) implies that $\|f_n - f\|_\infty \to 0$, so $f \in C_0(\bar{\Omega})$.

A similar argument leads to the following result.

Theorem 1.7.5. *Let Ω be bounded and convex with volume $|\Omega|$ and diameter β. If $p > N$ and $f \in W^{1,p}(\Omega)$ then f is continuous and bounded with*

$$\|f - f_\Omega\|_\infty \leq c \|\nabla f\|_p \frac{\beta^N}{|\Omega|} \beta^{1 - N/p}$$

where $c = c(N, p)$.

We now present some Sobolev inequalities for elliptic operators (of second order). Our treatment is far from complete, and is conditioned by the desire not to impose any smoothness conditions on the coefficients of the operators.

Theorem 1.7.6. *If H is strictly elliptic on $L^2(\Omega)$ where $\Omega \subseteq \mathbb{R}^N$ and $N \geq 3$, then there exists $c > 0$ such that*

$$|V| \leq c \|V\|_{N/2} H$$

for every potential $V \in L^{N/2}(\Omega)$.

Proof. By the strict ellipticity it is sufficient to prove that

$$\langle |V|f, f \rangle \leq c \|V\|_{N/2} \|\nabla f\|_2^2$$

for all $f \in C_c^\infty(\Omega)$. But

$$\langle |V|f, f \rangle \leq \|V\|_{N/2} \|f\|_q^2$$

where

$$1/q = 1/2 - 1/N$$

so we may apply Theorem 1.7.1.

The cases $N = 1$ and $N = 2$ need special treatment.

Theorem 1.7.7. *If H is strictly elliptic on $L^2(\Omega)$ where $\Omega \subseteq \mathbb{R}$ is an interval of length $\beta \leq \infty$, then there exist $c_i > 0$ such that*

$$|V| \leq c_1 \|V\|_1 \beta H \tag{1.7.3}$$

and

$$|V| \leq c_2 \|V\|_1 (\alpha H + \alpha^{-1}) \tag{1.7.4}$$

for all $\alpha > 0$ and all $V \in L^1(\Omega)$.

Proof. The proof of (1.7.3) follows that of Theorem 1.7.6, but uses Theorem 1.7.4 in place of Theorem 1.7.1. In order to prove (1.7.4) it is sufficient to treat the case $\Omega = \mathbb{R}$ and $H = -d^2/dx^2$. The bound is then equivalent to

$$\| |V|^{\frac{1}{2}} (\alpha H + \alpha^{-1})^{-\frac{1}{2}} \|^2 \leq c_2 \|V\|_1$$

which may be proved by computing the Hilbert–Schmidt norm of the operator using Fourier transforms.

Theorem 1.7.8. *Let H be strictly elliptic on $L^2(\Omega)$ where $\Omega \subseteq \mathbb{R}^2$, and let $V \in L^p(\Omega)$ for some $p > 1$. Then*

$$|V| \leq c_1 \|V\|_p (\alpha H + \alpha^{-1/(p-1)}) \tag{1.7.5}$$

for all $\alpha > 0$, where $c_1 = c_1(H, p)$. If Ω has finite diameter β then

$$|V| \leq c_2 \|V\|_p \beta^{2(p-1)/p} H \tag{1.7.6}$$

where $c_2 = c_2(H, p)$.

Proof. By the strict ellipticity it is sufficient to treat the case where $H = -\Delta$. For the proof of (1.7.5) we may also put $\Omega = \mathbb{R}^2$.

If $\mathscr{F}$ denotes the Fourier transform and
$$g(x) = \{\alpha x^2 + \alpha^{-1/(p-1)}\}^{\frac{1}{2}}$$
then
$$|V|^{\frac{1}{2}}\{\alpha H + \alpha^{-1/(p-1)}\}^{-\frac{1}{2}}f = |V|^{\frac{1}{2}}\mathscr{F}^{-1}\{g\mathscr{F}f\}$$
for all $f \in L^2$. We estimate the L^2 norm of this using Section 1.1. If
$$1/t = 1/2p + \tfrac{1}{2}$$
then $1 < t < 2$ and
$$\|g\mathscr{F}f\|_t \leq \|g\|_{2p}\|f\|_2.$$
If
$$1 = 1/s + 1/t$$
then $2 < s < \infty$ and
$$\|\mathscr{F}^{-1}\{g\mathscr{F}f\}\|_s \leq (2\pi)^{\frac{1}{2}-1/t}\|g\|_{2p}\|f\|_2.$$
But
$$\tfrac{1}{2} = 1/s + 1/2p$$
and $\|g\|_{2p}$ is independent of α, so
$$\||V|^{\frac{1}{2}}\mathscr{F}^{-1}(g\mathscr{F}f)\|_2 \leq \||V|^{\frac{1}{2}}\|_{2p}(2\pi)^{\frac{1}{2}-1/t}\|g\|_{2p}\|f\|_2$$
$$= c^{\frac{1}{2}}\|V\|_p^{\frac{1}{2}}\|f\|_2.$$
The operator bound
$$\||V|^{\frac{1}{2}}\{\alpha H + \alpha^{-1/(p-1)}\}^{-\frac{1}{2}}\| \leq c_1^{\frac{1}{2}}\|V\|_p^{\frac{1}{2}}$$
proved above is equivalent to (1.7.5).

We deduce (1.7.6) from (1.7.5) by using a ball B of radius β which contains Ω. If H' is $-\Delta$ on $L^2(B)$ with Dirichlet boundary conditions then
$$H \geq H' \geq c_3\beta^{-2}$$
by Theorem 1.5.8. Combining this with (1.7.5) leads to
$$|V| \leq c_1\|V\|_p(\alpha H + c_3^{-1}\beta^2\alpha^{-1/(p-1)}H)$$
$$\leq \frac{c_2}{2}\|V\|_p(\alpha + \beta^2\alpha^{-1/(p-1)})H.$$
We obtain (1.7.6) by putting $\alpha = \beta^{2(p-1)/p}$.

Although most of this book will concentrate on the study of elliptic differential operators with Dirichlet boundary conditions, we shall sometimes give analogous results in the case of Neumann boundary conditions. It turns out that Theorem 1.7.1 is not true in general if one replaces $W_0^{1,p}(\Omega)$ by $W^{1,p}(\Omega)$. We shall say that the region $\Omega \subseteq \mathbb{R}^N$ has the extension property if there exists a bounded linear map E from $W^{1,p}(\Omega)$ into $W^{1,p}(\mathbb{R}^N)$ such that

Some Sobolev inequalities

Ef is an extension of f from Ω to $\mathbb{R}^N$ for all $f \in W^{1,p}(\Omega)$ and all $1 \leq p \leq \infty$. We quote without proof the following fundamental result.

Proposition 1.7.9. *A region $\Omega \subseteq \mathbb{R}^N$ has the extension property if its boundary $\partial \Omega$ is minimally smooth in the following sense. There exists $\varepsilon > 0$, an integer k, an $M > 0$ and a possibly infinite sequence of open sets U_n such that:*
  (i) *if $x \in \partial \Omega$ then the ball with centre x and radius ε is contained in U_n for some n,*
  (ii) *no point in $\mathbb{R}^N$ is contained in more than k of the U_n,*
  (iii) *for each n there exists an isometry $T: \mathbb{R}^N \to \mathbb{R}^N$ and a function*

$$\phi_n: \mathbb{R}^{N-1} \to \mathbb{R}$$

such that

$$|\phi_n(x) - \phi_n(x')| \leq M|x - x'|$$

for all $x, x' \in \mathbb{R}^{N-1}$. Moreover

$$U_n \cap \Omega = U_n \cap T\Omega_n$$

where

$$\Omega_n = \{(x_1, \ldots, x_N): \phi_n(x_1, \ldots, x_{N-1}) < x_N\}.$$

Note 1.7.10. Every bounded convex region and every bounded region with piecewise smooth boundary has the extension property by the above proposition.

Lemma 1.7.11. *If the region Ω has the extension property, then the set $C_c^\infty(\Omega)$ of all restrictions to Ω of functions in $C_c^\infty(\mathbb{R}^N)$ is norm dense in $W^{1,p}(\Omega)$ for all $1 < p < \infty$. If also $1 \leq p < N$ and $1/q = 1/p - 1/N$ then there is a constant c_1 such that*

$$\|f\|_q \leq c_1(\|\nabla f\|_p^p + \|f\|_p^p)^{1/p}$$

for all $f \in W^{1,p}(\Omega)$.

Proof. If $f \in W^{1,p}(\Omega)$ and $\varepsilon > 0$ then there exists $g \in C_c^\infty(\mathbb{R}^N)$ such that

$$\int_{\mathbb{R}^N} \{|Ef - g|^p + |\nabla(Ef - g)|^p\} \, dx < \varepsilon^p$$

by Lemma 1.2.4 (with $\Omega = \mathbb{R}^N$). This implies

$$\int_\Omega \{|f - g|^p + |\nabla(f - g)|^p\} \, dx$$

$$= \int_\Omega \{|Ef - g|^p + |\nabla(Ef - g)|^p\} \, dx < \varepsilon^p$$

as required to prove the first statement.

By applying Theorem 1.7.1 (with $\Omega = \mathbb{R}^N$) we find that

$$\|f\|_q \leq \|Ef\|_q$$
$$\leq c\|\nabla(Ef)\|_p$$
$$\leq c\{\|\nabla(Ef)\|_p^p + \|Ef\|_p^p\}^{1/p}$$
$$\leq c\|E\|\{\|\nabla f\|_p^p + \|f\|_p^p\}^{1/p}$$

which proves the second statement.

We finally give a version of Theorem 1.6.8 for Neumann boundary conditions. It is of interest that some condition on Ω such as the extension property is actually necessary for the next result.

Theorem 1.7.12. *Let H be a uniformly elliptic operator on $L^2(\Omega)$ with Neumann boundary conditions. If Ω is bounded and has the extension property then H has compact resolvent.*

Proof. We follow the notation of Theorem 1.2.10, and write H_0 for $-\Delta$ on $L^2(\Omega)$ with Neumann boundary conditions. Since

$$0 \leq (H_0 + 1) \leq (1 + \lambda^{-1})(H + 1) = c^2(H + 1)$$

we see that

$$\|(H + 1)^{-\frac{1}{2}}(H_0 + 1)^{\frac{1}{2}}\|^2 = \|(H + 1)^{-\frac{1}{2}}(H_0 + 1)(H + 1)^{-\frac{1}{2}}\| \leq c^2.$$

Therefore

$$(H + 1)^{-1} = (H + 1)^{-\frac{1}{2}} \cdot (H + 1)^{-\frac{1}{2}}(H_0 + 1)^{\frac{1}{2}} \cdot (H_0 + 1)^{-\frac{1}{2}}$$

and the compactness of $(H + 1)^{-1}$ follows from that of $(H_0 + 1)^{\frac{1}{2}}$, or equivalently from that of $(H_0 + 1)^{-1}$. Also $(H_0 + 1)^{-1}$ is the adjoint of the identity map T from $W^{1,2}(\Omega)$ to $L^2(\Omega)$ and we shall show that $T = ADC$ where C is a bounded extension map from $W^{1,2}(\Omega)$ into $W_0^{1,2}(B)$ for some ball $B \supseteq \Omega$, D is the identity map from $W_0^{1,2}(B)$ to $L^2(B)$ and A is the restriction map from $L^2(B)$ to $L^2(\Omega)$. Since D^* is compact by Theorem 1.6.8, D is compact and T is compact.

The extension map C is defined as

$$(Cf)(x) = \phi(x)(Ef)(x)$$

where E is the extension map from $W^{1,2}(\Omega)$ into $W^{1,2}(\mathbb{R}^N)$ assumed to exist in the hypotheses, and ϕ is a positive C^∞ function which equals 1 on Ω and has compact support in B. The proof that C is bounded for the appropriate Sobolev norms is straightforward.

1.8. Definition of Schrödinger operators

We recall that an elliptic operator H_0 on $L^2(\Omega)$ with Dirichlet boundary

Definition of Schrödinger operators

conditions is defined by taking the quadratic form closure of

$$Q_0(f) = \int_\Omega \sum_{i,j} a_{ij}(x) \frac{\partial f}{\partial x_i} \frac{\partial \bar{f}}{\partial x_j} dx$$

where $f \in C_c^\infty(\Omega)$ and

$$\lambda(x)\mathbf{1} \leq a(x) \leq \mu(x)\mathbf{1}$$

on Ω for two positive continuous functions λ and μ on Ω. If V is a real-valued function in $L^1_{\text{loc}}(\Omega)$ (called a potential), then we define Q on $C_c^\infty(\Omega)$ by

$$Q(f) = Q_0(f) + \int_\Omega V(x)|f(x)|^2 dx$$

so that Q is associated with the formal expression

$$Lf = -\sum \frac{\partial}{\partial x_i}\left(a_{ij}\frac{\partial f}{\partial x_j}\right) + Vf$$

which we call a (generalised) Schrödinger operator.

Theorem 1.8.1. *If $0 \leq V \in L^1_{\text{loc}}(\Omega)$ then Q is closable on $C_c^\infty(\Omega)$ and its closure is the sum of the quadratic forms of H_0 and V, with*

$$\text{Dom}(Q) = \text{Quad}(H_0) \cap \text{Quad}(V). \tag{1.8.1}$$

The associated operator H on $L^2(\Omega)$ is the generator of a symmetric Markov semigroup.

Proof. Both H_0 and V are non-negative self-adjoint operators and their form sum H satisfies (1.8.1). By the Trotter product formula

$$e^{-Ht}f = \lim_{n \to \infty} (e^{-H_0 t/n} e^{-Vt/n})^n f \tag{1.8.2}$$

for all $t \geq 0$ and $f \in L^2(\Omega)$. It follows immediately from this formula that H satisfies Condition (iv) of Theorem 1.3.2 and Condition (i) of Theorem 1.3.3. We have therefore only to show that $C_c^\infty(\Omega)$ is a form core of H. We shall in fact show that if $0 \leq f \in \text{Quad}(H)$ then there exist $0 \leq f_n \in C_c^\infty(\Omega)$ with

$$\|f_n - f\|_2^2 + \|H_0^{\frac{1}{2}}(f - f_n)\|_2^2 + \|V^{\frac{1}{2}}(f - f_n)\|_2^2 \to 0 \tag{1.8.3}$$

as $n \to \infty$.

By Lemma 1.3.6 for such an f there exist $f_n \in W_c^{1,2}(\Omega)$ such that $0 \leq f_n \leq f$ and

$$\|f_n - f\|_2^2 + \|H_0^{\frac{1}{2}}(f - f_n)\|_2^2 \to 0.$$

Since $0 \leq V|f|^2 \in L^1(\Omega)$ one can deduce from the dominated convergence

theorem that
$$\|V^{\frac{1}{2}}(f-f_n)\|_2^2 \to 0.$$
It is therefore sufficient to treat the case where
$$0 \leqslant f \in W_c^{1,2}(\Omega) \cap \text{Quad}(H_0) \cap \text{Quad}(V).$$
For such an f we put $f_n = f \wedge n$ and note that
$$\|H_0^{\frac{1}{2}} f_n\|_2^2 \leqslant \|H_0^{\frac{1}{2}} f\|_2^2$$
by Theorem 1.3.3, while
$$\|V^{\frac{1}{2}} f_n\|_2^2 \leqslant \|V^{\frac{1}{2}} f\|_2^2$$
by pointwise domination. This implies that (1.8.3) holds, and hence that it is sufficient to treat the case where
$$0 \leqslant f \in W_c^{1,2}(\Omega) \cap L_c^\infty(\Omega) \subseteq \text{Quad}(H_0) \cap \text{Quad}(V).$$
This final case is handled by a standard mollifier argument as in Lemma 1.2.4.

The above results are easily modified to deal with $V \in L^1_{\text{loc}}(\Omega)$ which are bounded below, and there are many methods of controlling potentials which are more singular.

Theorem 1.8.2. *Let H_0 be an elliptic operator satisfying Dirichlet boundary conditions on $L^2(\Omega)$, where Ω is a region in $\mathbb{R}^N$. If $V \in L^1_{\text{loc}}(\Omega)$ satisfies*
$$V_- \leqslant \delta H_0 + \gamma \tag{1.8.4}$$
in the sense of quadratic forms for some $0 < \delta < 1$ and some $\gamma \in \mathbb{R}$, then the form sum H of H_0 and V is self-adjoint and bounded below with
$$\text{Quad}(H) = \text{Quad}(H_0) \cap \text{Quad}(V_+).$$
Moreover e^{-Ht} is a positivity-preserving semigroup on $L^2(\Omega)$.

Proof. We first define H_1 to be the form sum of H_0 and V_+ as in Theorem 1.8.1 and note that
$$V_- \leqslant \delta H_1 + \gamma.$$
By a standard perturbation lemma one may now define H to be the form sum of H_1 and $-V_-$ to obtain
$$\text{Quad}(H) = \text{Quad}(H_1).$$
Finally one sees that $H_n = H_0 + V_n$ converges in the sense of quadratic forms to H, where $V_n = V \vee (-n)$. But $e^{-(H_n + n)t}$ is positivity-preserving by Theorem 1.8.1. If $0 \leqslant f \in L^2(\Omega)$ we deduce that
$$e^{-Ht} f = \lim_{n \to \infty} e^{-H_n t} f \geqslant 0.$$

Definition of Schrödinger operators

One way of verifying (1.8.4) for locally bounded potentials which become very negative as one approaches $\partial\Omega$ is by applying the ideas in Section 1.5. Local singularities of V_- can, however, often be dealt with by the following result, which can be modified to treat the cases $N = 1, 2$ by using results in Section 1.7.

Theorem 1.8.3. *Let Ω be a region in $\mathbb{R}^N$ and let H_0 be strictly elliptic in $L^2(\Omega)$. If $N \geq 3$ and $W \in L^{N/2}(\Omega)$ then for all $0 < \delta < \infty$ there exists $\gamma(\delta) < \infty$ such that*

$$|W| \leq \delta H_0 + \gamma(\delta)$$

in the sense of quadratic forms.

Proof. We recall from Theorem 1.7.6 that there is a constant c such that

$$|V| \leq c \|V\|_{N/2} H_0$$

for any potential V. Also we may write $W = V + X$ where $\|V\|_{N/2} \leq \delta/c$ and X is bounded. We then obtain

$$|W| \leq |V| + \|X\|_\infty$$
$$\leq \delta H_0 + \|X\|_\infty.$$

One defect of the above theorem is that one cannot specify the function γ in terms of $\|W\|_{N/2}$ alone. The next result overcomes this problem at the cost of a slightly stronger hypothesis.

Theorem 1.8.4. *If in the last theorem one assumes instead that $W \in L^p(\Omega)$ for some $p > N/2$, then one has*

$$|W| \leq \delta H_0 + k\delta^{-N/(2p-N)}$$

for all $0 < \delta < \infty$, where $k < \infty$ depends upon $N, p, \|W\|_p$ and the ellipticity constant of H_0.

Proof. We note that $|W|^\alpha \in L^{N/2}(\Omega)$ if $\alpha = 2p/N > 1$ and that

$$s \leq \varepsilon s^\alpha + (\alpha\varepsilon)^{-1/(\alpha-1)}$$

for all $s \geq 0$ and $\varepsilon > 0$. Therefore

$$|W| \leq \varepsilon |W|^\alpha + (\alpha\varepsilon)^{-1/(\alpha-1)}$$
$$\leq \varepsilon c \| |W|^\alpha \|_{N/2} H_0 + (\alpha\varepsilon)^{-1/(\alpha-1)}$$
$$= \varepsilon c \|W\|_p^\alpha H_0 + (\alpha\varepsilon)^{-1/(\alpha-1)}.$$

The proof is completed by putting $\delta = \varepsilon c \|W\|_p^\alpha$.

There are many more sophisticated hypotheses on W which yield much the same results as the above; for example we could assume that W is uniformly locally L^p for some $p > N/2$ or that W lies in the class K_N. However all such variations must accommodate the fact that the potential $W(x) = |x|^{-\alpha}$ is form bounded with respect to $-\Delta$ on $L^2(\mathbb{R}^3)$ if and only if $\alpha \leqslant 2$, and that the form sum

$$H = -\Delta + c|x|^{-2}$$

is bounded below if and only if $c \geqslant -\frac{1}{4}$.

1.9 The asymptotic eigenvalue distribution

If Ω is a region in $\mathbb{R}^N$ and H is $-\Delta$ on $L^2(\Omega)$ subject to Dirichlet boundary conditions then e^{-Ht} has a positive C^∞ integral kernel K which satisfies

$$0 < K(t, x, y) \leqslant (4\pi t)^{-N/2} e^{-(x-y)^2/4t} \qquad (1.9.1)$$

for all $t > 0$ and $x, y \in \Omega$. This follows immediately from the Feynman–Kac formula, but may also be deduced from Theorem 5.2.1 and Example 2.1.8 below. If Ω has finite volume, then it follows from (1.9.1) that e^{-Ht} is trace class with

$$0 < \operatorname{tr}[e^{-Ht}]$$
$$= \int_\Omega K(t, x, x)\, dx$$
$$\leqslant (4\pi t)^{-N/2} |\Omega|.$$

One of the earliest results in spectral theory was Weyl's theorem, which determined the asymptotic distribution of the eigenvalues E_n of Ω. We shall concentrate instead on the closely related question of the asymptotic behaviour of

$$\operatorname{tr}[e^{-Ht}] = \sum_{n=0}^\infty e^{-E_n t}$$

as $t \to 0$. Weyl's theorem then states that

$$\lim_{t \to 0} \operatorname{tr}[e^{-Ht}](4\pi t)^{N/2} = |\Omega|$$

provided the RHS is finite. This theorem has been subsequently generalised to yield an asymptotic expansion

$$\operatorname{tr}[e^{-Ht}] = (4\pi t)^{-N/2}\{|\Omega| - \tfrac{1}{2}\pi^{\frac{1}{2}} t^{\frac{1}{2}} |\partial\Omega| + O(t)\}$$

as $t \to 0$. This subject has been developed to a high degree of sophistication, and we do not pursue it here.

The above theory is not applicable to a class of regions of infinite volume for which $\operatorname{tr}[e^{-Ht}]$ is nevertheless finite for all $t > 0$. The asymptotic eigenvalue distribution of the relevant operators is quite different and can be investigated using the ideas of Section 1.5. Throughout this section we shall assume that Ω is a strongly regular region so that

$$H \geqslant c/d^2$$

for some $c > 0$, where d is the distance function, and we shall assume that $d(x) \to 0$ as $|x| \to \infty$ so that $E_0 > 0$ by Theorem 1.5.8 and H^{-1} is compact by Theorem 1.6.8. For a typical case of the regions which we shall study see Example 1.6.9.

Theorem 1.9.1. *If Ω is strongly regular then there exists $c > 0$ such that*

$$0 < \mathrm{tr}\,[e^{-Ht}] \leqslant (2\pi t)^{-N/2} \int_\Omega e^{-ct/2d^2}$$

whenever the RHS is finite.

Proof. The Golden–Thompson inequality states that if $A \geqslant 0$ and $B \geqslant 0$ are possibly unbounded self-adjoint operators then

$$0 \leqslant \mathrm{tr}\,[e^{-(A+B)}] \leqslant \mathrm{tr}\,[e^{-A/2} e^{-B} e^{-A/2}] \qquad (1.9.2)$$

whenever the RHS is finite. Now

$$H \geqslant \tfrac{1}{2}H + c/2d^2$$

so

$$\begin{aligned}
\mathrm{tr}\,[e^{-Ht}] &\leqslant \mathrm{tr}\,[e^{(\frac{1}{2}H + c/2d^2)t}] \\
&\leqslant \mathrm{tr}\,[e^{-ct/4d^2} e^{-Ht/2} e^{-ct/4d^2}] \\
&= \int_\Omega e^{-ct/2d^2} K(t/2, x, x)\,dx \\
&\leqslant (2\pi t)^{-N/2} \int_\Omega e^{-ct/2d^2}
\end{aligned}$$

by (1.9.1).

Our lower bound on $\mathrm{tr}\,[e^{-Ht}]$ uses the standard technique of decomposing Ω into subregions by introducing Dirichlet boundary conditions along various surfaces. We say that a set $C \subseteq \mathbb{R}^n$ is a standard cube if

$$C = \{x \in \mathbb{R}^N : a_i 2^{-n} < x_i < (a_i + 1) 2^{-n}\}$$

for some $a \in \mathbb{Z}^N$ and some $n \in \mathbb{Z}$. We let $\{C_n\}_{n=1}^\infty$ denote the set of standard cubes which are contained in Ω and which are maximal (under the inclusion ordering) among all such cubes and let δ_n be the edge length of C_n.

Lemma 1.9.2. *If d is bounded on Ω then $C_m \cap C_n = \emptyset$ for $m \neq n$ and*

$$\bigcup_n C_n \subseteq \Omega \subseteq \bigcup_n \bar{C}_n.$$

Moreover if $x \in \bar{C}_n$ then

$$d(x) \leq 2N^{\frac{1}{2}}\delta_n.$$

Proof. This is a matter of simple geometry.

Theorem 1.9.3. *If $0 < t < \infty$ then*

$$(8\pi t)^{-N/2} \int_\Omega e^{-8\pi^2 Nt/d^2} \leq \operatorname{tr}[e^{-Ht}].$$

Proof. Let H' denote $-\Delta$ on $\Omega' = \bigcup_n C_n$, K_n denote $-\Delta$ on C_n and K denote $-\Delta$ on the unit cube $C \subseteq \mathbb{R}^N$, all subject to Dirichlet boundary conditions. Then $\Omega' \subseteq \Omega$ implies $H' \geq H$, and it follows from Section 1.1.11 that

$$\operatorname{tr}[e^{-Ht}] \geq \operatorname{tr}[e^{-H't}] = \sum_{n=1}^\infty \operatorname{tr}[\exp(-K_n t)]$$

$$= \sum_{n=1}^\infty \operatorname{tr}[\exp(-K\delta_n^{-2}t)].$$

Moreover

$$\{\operatorname{tr}[e^{-Ks}]\}^{1/N} = \sum_{n=0}^\infty e^{-\pi^2(n+1)^2 s}$$

$$\geq \sum_{n=0}^\infty e^{-2\pi^2(n^2+1)s}$$

$$\geq \tfrac{1}{2} e^{-2\pi^2 s} \sum_{n=-\infty}^\infty e^{-2\pi^2 n^2 s}$$

$$= e^{-2\pi^2 s}(8\pi s)^{-\frac{1}{2}} \sum_{n=-\infty}^\infty e^{-n^2/2s}$$

$$\geq e^{-2\pi^2 s}(8\pi s)^{-\frac{1}{2}}$$

by the Poisson summation formula. Therefore

$$\operatorname{tr}[e^{-Ht}] \geq (8\pi t)^{-N/2} \sum_{n=0}^\infty \delta_n^N \exp(-2\pi^2 N\delta_n^{-2} t)$$

$$\geq (8\pi t)^{-N/2} \sum_{n=0}^\infty \int_{C_n} \exp(-8\pi^2 N^2 t/d^2)$$

$$= (8\pi t)^{-N/2} \int_\Omega \exp(-8\pi^2 N^2 t/d^2).$$

Corollary 1.9.4. *If U is strongly regular and d is bounded then*

$$(8\pi t)^{-N/2} \int_\Omega e^{-8\pi^2 N^2 t/d^2} \leq \operatorname{tr}[e^{-Ht}]$$

$$\leq (2\pi t)^{-N/2} \int_\Omega e^{-ct/2d^2}$$

for all $0 < t < \infty$. Hence $\text{tr}[e^{-Ht}] < \infty$ for all $0 < t < \infty$ if and only if

$$\int_\Omega e^{-s/d^2} < \infty$$

for all $0 < s < \infty$.

Example 1.9.5. Define $\Omega_\gamma \subseteq \mathbb{R}^2$ by

$$\Omega_\gamma = \{(x, y) : x > 1 \text{ and } |y| < x^{-\gamma}\}$$

where $0 < \gamma < 1$. Then Ω_γ has infinite area so Weyl's theorem is inapplicable. However

$$\int_{\Omega_\gamma} e^{-t/d^2} \sim \int_1^\infty x^{-\gamma} \exp(-tx^{2\gamma}) \, dx$$

$$\sim t^{-(1/2\gamma - 1/2)}$$

as $t \to 0$. Therefore

$$\text{tr}[e^{-Ht}] \sim t^{-(1/2\gamma + 1/2)}$$

as $t \to 0$ by Corollary 1.9.4. The above method does not however allow us to prove the existence of

$$\lim_{t \to 0} t^{(1/2\gamma + 1/2)} \text{tr}[e^{-Ht}].$$

Theorem 1.9.6. Suppose that Ω is strongly regular, d is bounded and $\gamma > N/2$. Then $\text{tr}[H^{-\gamma}]$ is finite if and only if

$$\int_\Omega d^{(2\gamma - N)} < \infty$$

Proof. We have

$$\text{tr}[H^{-\gamma}] = \Gamma(\gamma)^{-1} \int_0^\infty t^{\gamma - 1} \text{tr}[e^{-Ht}] \, dt$$

so by Corollary 1.9.4 the result follows from the observation that

$$\int_\Omega \int_0^\infty t^{-N/2} e^{-at/d^2} t^{\gamma - 1} \, dt = \Gamma(\gamma - N/2) \int_\Omega (d^2/a)^{\gamma - N/2}$$

for any $a > 0$.

Notes

Section 1.1 Most of this standard material may be found in Dunford and Schwartz (1958) or Reed and Simon (1975). We refer to Stein and Weiss

(1971) for Section 1.1.6, Reed and Simon (1978) for Sections 1.1.10 and 1.1.11, and Stein (1970) for Section 1.1.12.

Section 1.2 An exposition of the abstract theory of quadratic forms may be found in Davies (1980) and its application to second order elliptic operators is treated in detail by Fukushima (1980). The conditions in Theorems 1.2.5 and 1.2.6 for the quadratic form Q to be closable on $C_c^\infty(\Omega)$ are far from the best possible; see Röckner and Wielens (1985) for much sharper results. The operator H defined by these theorems is not generally essentially self-adjoint on $C_c^\infty(\Omega)$ except in one dimension. For a variety of criteria for self-adjointness see Chernoff (1973), Cordes (1987) p. 108, Davies (1985D), Frehse (1977), Kalf and Walter (1972), Kato (1981), Strichartz (1983). A proof of Lemma 1.2.9 may be found in Adams (1975) p. 52.

Section 1.3 Most of this material may be found in Davies (1980), Fukushima (1980) or Reed and Simon (1978). Quadratic forms satisfying the conditions of Theorems 1.3.2 and 1.3.3 are called Dirichlet forms; a partial classification of such forms, due to Beurling and Deny, may be found in Fukushima (1980), p. 48. For a deeper study of the construction of continuous time Markov chains than that of Theorem 1.3.10 see Rogers and Williams (1986). Our definition of the Laplacian on a graph in Example 1.3.11 is taken from Dodziuk and Kendall (1986), who give many further results. For applications of the ideas in Theorem 1.3.10 and Example 1.3.11 to random walks on finitely generated groups and hence to Riemannian manifolds see Brooks (1981B; 1985), Lyons and Sullivan (1984) and Varopoulos (1984; 1985A; 1985B; 1986).

Section 1.4 Fukushima (1980) and Silverstein (1974) are general references for this material, and in particular for the construction of the Hunt process associated with a Dirichlet form. Theorem 1.4.2 is taken from Reed and Simon (1975) p. 255. A proof of Proposition 1.4.3 may be found in Davies (1980) p. 174. The origins of Theorem 1.4.4 on the conservation of probability go back to Hasminskii; see Davies (1985D) for a recent survey.

Section 1.5 A version of Lemma 1.5.1 may be found in Hardy, Littlewood and Polya (1952) p. 175; for its relationship with the uncertainty principle see Faris (1978). Theorem 1.5.3 is due to Davies (1984); an analogous theorem for the Laplace operator on a Riemannian manifold was obtained by Croke and Derdzinski (1987) by integrating over the geodesic flow. Theorem 1.5.4 appears with slight variations in Davies (1984) and Ancona (1986), but for smooth bounded domains goes back to Kadlec and Kufner (1966). Many analogues of these theorems for more general elliptic operators on nice domains have been proved; see Kalf and Walter (1972),

Lions and Magenes (1972) p. 69 and Davies (1985B). A version of Theorem 1.5.6 for higher order Sobolev spaces on smooth bounded domains was proved by Kadlec and Kufner (1966) and D. J. Harris; see Edmunds and Evans (1987) Section V.3 for further details. Results along the lines of Theorems 1.5.8 and 1.5.10 were first proved by Hayman (1977/8) with improvements and simplifications described in Osserman (1977; 1980). Our proof of Theorem 1.5.10 follows Ancona (1986); see Hille (1962) or Duren (1983) for proofs of the Koebe $\frac{1}{4}$ theorem. Theorem 1.5.11 was proved along with many related results by Lieb (1983), who used a different method from that presented here. Theorem 1.5.12 is a classical result going back to Jacobi. It has a converse due to Allegretto (1974) and Moss and Piepenbrink (1978); for applications see Agmon (1982) and Simon (1982). For other material related to Theorem 1.5.14 see the survey of Davies (1985D).

Section 1.6 For Theorem 1.6.1 see Persson (1964) or Bergh and Löfström (1976) p. 85. One may enquire whether the spectrum of an elliptic operator on L^p is independent of p if one does not make the compactness assumption of Theorem 1.6.3. This was proved for a very wide class of Schrödinger operators on Euclidean space by Hempel and Voigt (1986; 1987); however Davies, Simon and Taylor (1988) showed that one does have p-dependence for the Laplacian on hyperbolic space and for many Kleinian groups. Theorems 1.6.6 and 1.6.7 are taken from Davies (1985D); for substantial extensions of these results see Pang (1987). Necessary and sufficient conditions on $\Omega \subseteq \mathbb{R}^N$ for Δ^{-1} to have compact resolvent were found by Molchanov (1953); for recent accounts see Adams (1975), Maz'ja (1979) and Edmunds and Evans (1987). Molchanov's conditions involve capacity considerations, and are quite different from those of Theorem 1.6.8, due to Davies (1985A).

Section 1.7 These bounds are classical and are to be found in Adams (1975) or Gilbarg and Trudinger (1977). For Proposition 1.7.9 see Stein (1970) p. 181.

Section 1.8 For a proof of the Trotter product formula (1.8.2) see Kato (1978) or Davies (1980). Theorems 1.8.2, 1.8.3 and 1.8.4 are classical and may be found in Reed and Simon (1975). Simon (1979) describes the approach to Schrödinger operators using functional integration, and Simon (1982) surveys the properties of potentials in the class K_N.

Section 1.9 The asymptotic formula of Weyl (1912) for tr $[e^{-Ht}]$ is extended to a wide class of elliptic operators in Courant and Hilbert (1953). It is the leading term of an asymptotic expansion analysed by McKean and Singer

(1967) and others in great detail. For the Laplacian on a region in Euclidean space sharp estimates of the error terms in the asymptotic expansion have been given by van den Berg (1984A; 1987A). However, the theory of this section all comes from Davies (1985A). For references to various forms of the Golden–Thompson inequality see Reed and Simon (1975) p. 333. For certain regions of infinite volume, such as that of Example 1.9.5, the leading order coefficient has been calculated; see Rosenbljum (1972; 1973), Tamura (1976), Fleckinger (1981), Simon (1983A; 1983B) and van den Berg (1984B). However, an asymptotically exact formula for the leading order of the trace as $t \to 0$ is not known for general regions.

2
Logarithmic Sobolev inequalities

2.1 Contractivity properties

Let e^{-Ht} be a symmetric Markov semigroup on $L^2(\Omega, dx)$ where dx is a Borel measure on the locally compact, second countable Hausdorff space Ω. The main goal of this chapter is to use regularising properties of e^{-Ht} to obtain bounds on the integral kernel $K(t, x, y)$ of e^{-Ht}, which we shall call the heat kernel.

As a first example of such properties we say that e^{-Ht} is hypercontractive if there exists $t > 0$ such that e^{-Ht} is bounded from L^2 to L^4. We quote the following theorem.

Theorem 2.1.1. *If e^{-Ht} is hypercontractive and $1 < p \leq q < \infty$, then there exists $T(p, q) < \infty$ such that e^{-Ht} is a contraction from L^p to L^q for $t \geq T(p, q)$.*

Although hypercontractivity has been of great importance in quantum field theory, an even stronger property often holds for elliptic differential operators, and it is this property on which we concentrate.

We say that e^{-Ht} is ultracontractive if e^{-Ht} is bounded from L^2 to L^∞ for all $t > 0$. If we define $\|A\|_{q,p}$ to be the norm of an operator A from L^p to L^q, then it is clear that

$$c_t = \|e^{-Ht}\|_{\infty, 2} \tag{2.1.1}$$

is a monotonically decreasing function of t. Moreover $c_t \to +\infty$ as $t \to 0$ unless $L^2 \subseteq L^\infty$, which implies that dx is a purely atomic measure. See Theorem 1.3.10 for an example of a random walk for which one has $0 \leq c_t \leq 1$ for all $t \geq 0$.

Lemma 2.1.2. *If e^{-Ht} is ultracontractive then it has an integral kernel $K(t, x, y)$ for all $t > 0$, which satisfies*

$$0 \leq K(t, x, y) \leq c_{t/2}^2$$

almost everywhere. Conversely if e^{-Ht} has a kernel satisfying

$$0 \leq K(t, x, y) \leq a_t < \infty \tag{2.1.2}$$

almost everywhere then e^{-Ht} is ultracontractive with

$$c_t \leqslant a_t^{\frac{1}{2}}. \qquad (2.1.3)$$

Proof. By taking adjoints we see that (2.1.1) implies

$$c_t = \|e^{-Ht}\|_{2,1}$$

so

$$\|e^{-Ht}\|_{\infty,1} \leqslant \|e^{-Ht/2}\|_{\infty,2} \|e^{-Ht/2}\|_{2,1}$$
$$= c_{t/2}^2.$$

We now use the theorem that every bounded operator from L^1 to L^∞ has a kernel whose L^∞ norm equals the operator norm.

For the converse we observe that (2.1.2) implies easily that

$$\|e^{-Ht}\|_{\infty,1} \leqslant a_t.$$

We obtain (2.1.3) by interpolating between this and the bound

$$\|e^{-Ht}\|_{\infty,\infty} \leqslant 1.$$

Example 2.1.3. Let $H = -\Delta$ on $L^2(\mathbb{R}^N, dx)$. Then we stated in Section 1.8 that

$$K(t, x, y) = K_t(x - y)$$

where

$$K_t(x) = (4\pi t)^{-N/2} e^{-x^2/4t}.$$

It is immediate that we have ultracontractivity with

$$c_t = \|K_t\|_2 = (8\pi t)^{-N/4}$$

and

$$a_t = \|K_t\|_\infty = (4\pi t)^{-N/2}.$$

Our ultimate goal will be to obtain upper and lower bounds on the heat kernels of general elliptic operators which are as close as possible to those of the above example. In this section we investigate some spectral consequences of ultracontractivity.

Theorem 2.1.4. Let e^{-Ht} be ultracontractive where Ω has finite measure $|\Omega|$. Then

$$\operatorname{tr}[e^{-Ht}] < \infty$$

for all $t > 0$. Let $\{E_n\}_{n=0}^\infty$ be the eigenvalues of H written in increasing order and repeated according to multiplicity, and let ϕ_n be the corresponding eigenfunctions normalised by $\|\phi_n\|_2 = 1$. Then $\phi_n \in L^\infty$ for all n and we have

$$K(t, x, y) = \sum_{n=0}^\infty \exp(-E_n t) \phi_n(x) \phi_n(y) \qquad (2.1.4)$$

where the series converges uniformly on $[\alpha, \infty) \times \Omega \times \Omega$ for all $\alpha > 0$.

Proof. If η_t denotes the Hilbert–Schmidt norm of e^{-Ht} then

$$\operatorname{tr}[e^{-Ht}] = \eta_{t/2}^2$$
$$= \int_{\Omega \times \Omega} K(t/2, x, y)^2 \, dx \, dy$$
$$\leqslant c_{t/4}^4 |\Omega|^2 < \infty$$

by Lemma 2.1.2. Since the identity (2.1.4) certainly holds if the convergence is computed in $L^2(\Omega \times \Omega)$, we have only to check that the RHS is uniformly convergent.

From the bound

$$\|\exp(-E_n t/3)\phi_n\|_\infty = \|\exp(-Ht/3)\phi_n\|_\infty \leqslant c_{t/3}$$

we see that

$$|\exp(-E_n t)\phi_n(x)\phi_n(y)| \leqslant \exp(-E_n t/3) c_{t/3}^2$$

so the RHS of (2.1.4) is dominated by

$$\sum_{n=0}^\infty \exp(-E_n \alpha/3) c_{\alpha/3}^2 = \operatorname{tr}[\exp(-H\alpha/3)] c_{\alpha/3}^2 < \infty$$

and the Weierstrass M-test is applicable.

Theorem 2.1.5. *If e^{-Ht} is ultracontractive and $|\Omega| < \infty$ then e^{-Ht} is compact on L^p for all $1 \leqslant p \leqslant \infty$ and $0 < t < \infty$. The spectrum of H_p is independent of p for $1 \leqslant p \leqslant \infty$, and e^{-Ht} is norm analytic on L^p for $1 \leqslant p \leqslant \infty$ and $0 < t < \infty$.*

Proof. The compactness of e^{-Ht} on L^1 follows from the identity

$$e^{-Ht} = ABC$$

where $C = e^{-Ht/2}$ is bounded from L^1 to L^2, $B = e^{-Ht/2}$ is compact from L^2 to L^2, and $A = 1$ is bounded from L^2 to L^1. For the second statement of the theorem we quote Theorem 1.6.4.

By duality it is sufficient to prove analyticity when $1 \leqslant p \leqslant 2$ and $\operatorname{Re} t > \alpha$, where $\alpha > 0$ is arbitrary. This follows from the formula

$$e^{-Ht} = AB_t C$$

where $C = e^{-H\alpha}$ is bounded from L^p to L^2 by ultracontractivity and interpolation, $B_t = e^{-H(t-\alpha)}$ is analytic from L^2 to L^2 by the spectral theorem, and $A = 1$ is bounded from L^2 to L^p.

The application of the above theorems to elliptic differential operators is aided by the following theorems. Let H_Ω be an elliptic differential operator on $L^2(\Omega)$ with associated form Q, where Ω in a region is $\mathbb{R}^N$ and we assume that Dirichlet boundary conditions are imposed in the usual manner. Let $\Sigma \subseteq \Omega$ and let H_Σ be the restriction of H_Ω to $L^2(\Sigma)$ in the following sense. We define $\text{Dom}(Q_\Sigma)$ to be the completion on $C_c^\infty(\Sigma)$ in $\text{Dom}(Q_\Omega)$ for the norm associated with (1.2.2). We then put

$$Q_\Sigma(f) = \begin{cases} Q_\Omega(f) & \text{if } f \in \text{Dom}(Q_\Sigma) \\ +\infty & \text{otherwise.} \end{cases}$$

Theorem 2.1.6. *If H_Ω and H_Σ are defined as above and $\exp(-H_\Sigma t)$ is extended from $L^2(\Sigma)$ to $L^2(\Omega)$ by putting $\exp(-H_\Sigma t)f = 0$ for all $f \in L^2(\Omega \backslash \Sigma)$, then*

$$0 \leqslant \exp(-H_\Sigma t)f \leqslant \exp(-H_\Omega t)f \tag{2.1.5}$$

for all $0 \leqslant f \in L^2(\Omega)$. If the two semigroups possess finite heat kernels then

$$0 \leqslant K_\Sigma(t, x, y) \leqslant K_\Omega(t, x, y). \tag{2.1.6}$$

Proof. Let Σ_n be a sequence of open subsets of Σ such that Σ_n has compact closure and $\bar\Sigma_n \subseteq \Sigma_{n+1}$ for all n and $\bigcup_n \Sigma_n = \Sigma$. Also let χ_n denote the characteristic function of $\Omega \backslash \Sigma_n$. Then the operators $(H_\Omega + m\chi_n)$ increase monotonically as $m \to \infty$, and the forms converge in the sense of Theorem 1.2.2 to a form Q_n which is a restriction of Q_Σ. The forms Q_n decrease as $n \to \infty$ and converge to Q_Σ in the sense of Theorem 1.2.3 because $C_c^\infty(\Sigma)$ is a core of Q_Σ and every $f \in C_c^\infty(\Sigma)$ lies in $\text{Dom}(Q_n)$ for large enough n. Now it is evident from perturbation theory or the Trotter product formula that

$$0 \leqslant \exp(-(H_\Omega + m\chi_n)t)f \leqslant \exp(-H_\Omega t)f$$

for all $0 \leqslant f \in L^2(\Omega)$, so (2.1.5) follows by taking the two limits. The equivalence of (2.1.5) and (2.1.6) is standard.

Theorem 2.1.7. *If H_Ω and H_Σ are as in Theorem 2.1.6 and $\exp(-H_\Omega t)$ is ultracontractive, then $\exp(-H_\Sigma t)$ is ultracontractive.*

Proof. If $f \in L^2(\Omega)$ then we have

$$|\exp(-H_\Sigma t)f| \leqslant \exp(-H_\Sigma t)|f| \leqslant \exp(-H_\Omega t)|f|$$

so

$$\|\exp(-H_\Sigma t)f\|_\infty \leqslant \|\exp(-H_\Omega t)|f|\|_\infty \leqslant c_\Omega(t)\|f\|_2.$$

Example 2.1.8. Let Ω be a region in $\mathbb{R}^N$ and let $H = -\Delta$ on $L^2(\Omega)$ with Dirichlet boundary conditions. By Example 2.1.3 and Theorem 2.1.7 we see that e^{-Ht} is ultracontractive with a kernel which satisfies

$$0 \leqslant K_0(t, x, y) \leqslant (4\pi t)^{-N/2} e^{-(x-y)^2/4t}.$$

One expects K_0 to vanish as x or y approach the boundary $\partial\Omega$, but this depends upon some boundary regularity conditions. If Ω is bounded then H has discrete spectrum by Theorem 2.1.4. If $H\phi_n = E_n\phi_n$ and $\|\phi_n\|_2 = 1$ then ϕ_n is bounded and

$$\|\phi_n\|_\infty \leqslant \exp(E_n t)c_t$$
$$\leqslant \exp(E_n t)(8\pi t)^{-N/4}.$$

Putting $8\pi t = E_n^{-1}$ we obtain the bound

$$\|\phi_n\|_\infty \leqslant e^{1/8\pi} E_n^{N/4}.$$

Upper and lower bounds on E_n may be obtained by minimax.

Example 2.1.9. Let Ω be a region in $\mathbb{R}^N$ and let $H = -\Delta + V$ where $0 \leqslant V \in L^1_{\text{loc}}(\Omega)$ and the Schrödinger operator H is defined as explained in Section 1.8. Then (1.8.2) implies

$$0 \leqslant K(t, x, y) \leqslant K_0(t, x, y)$$
$$\leqslant (4\pi t)^{-N/2} e^{-(x-y)^2/4t}$$

where K_0 is as in Example 2.1.8. Therefore e^{-Ht} is a symmetric Markov semigroup on $L^2(\Omega)$ and is ultracontractive with

$$\|e^{-Ht}\|_{\infty,1} \leqslant (4\pi t)^{-N/2}$$

for all $0 < t < \infty$.

If V is not bounded below then the proof that e^{-Ht} is still ultracontractive under suitable hypotheses is much harder. See Section 4.8 for details.

2.2 The fundamental inequalities

The key to all our subsequent theory will be Theorems 2.2.3 and 2.2.7. Conceptually both theorems are extremely simple, relying solely upon the solution of certain first order differential inequalities. Unfortunately a little care is needed to prove that the formal steps carried out are justified, but the technicalities involved are of a routine nature. *We assume throughout that* e^{-Ht} *is a symmetric Markov semigroup on* $L^2(\Omega, dx)$ *where* dx *is a Borel measure on a locally compact, second countable, Hausdorff space* Ω.

Lemma 2.2.1. *Let* $0 \leqslant f \in \text{Dom}(H) \cap L^1 \cap L^\infty$ *and put* $f_s = e^{-Hs} f$. *Let* $p(s)$ *be a continuously differentiable function from* $[0, t)$ *into* $[2, \infty)$. *Then* $f'_s f_s^{p(s)-1}$ *and* $f_s^{p(s)} \log f_s$ *are norm continuous functions of* $s \in [0, t)$ *with values in* L^1.

Proof. For the first part we note that f'_s is a norm continuous function of s with values in L^2, so we have only to prove the same for $f_s^{p(s)-1}$. If $0 < \lambda < t$

then $f_u(x)$, $f_v(x)$, $p(u)$ and $p(v)$ are uniformly bounded for $u, v \in [0, \lambda]$ and $x \in \Omega$, so

$$|f_u^{p(u)-1}(x) - f_v^{p(v)-1}(x)| \leq |f_u^{p(u)-1}(x) - f_v^{p(u)-1}(x)| + |f_v^{p(u)-1} - f_v^{p(v)-1}(x)|$$
$$\leq c_1 |f_u(x) - f_v(x)| + c_2 |p(u) - p(v)| |f_v(x)|^{\frac{1}{2}}.$$

This implies that

$$\|f_u^{p(u)-1} - f_v^{p(v)-1}\|_2 \leq c_1 \|f_u - f_v\|_2 + c_2 |p(u) - p(v)| \|f_v\|_1^{\frac{1}{2}}$$
$$\leq c_1 \|f_u - f_v\|_2 + c_2 |p(u) - p(v)| \|f_0\|_1^{\frac{1}{2}}$$

as required.

For the second part we let λ, u, v be as above and observe that

$$|f_u^{p(u)}(x) \log f_u(x) - f_v^{p(v)}(x) \log f_v(x)|$$
$$\leq |f_u^{p(u)}(x) \log f_u(x) - f_u^{p(v)}(x) \log f_u(x)|$$
$$+ |f_u^{p(v)}(x) \log f_u(x) - f_v^{p(v)}(x) \log f_v(x)|$$
$$\leq c_1 |p(u) - p(v)| |f_u(x)| + c_2 |f_u(x) - f_v(x)|.$$

Therefore

$$\|f_u^{p(u)} \log f_u - f_v^{p(v)} \log f_v\|_1 \leq c_1 |p(u) - p(v)| \|f_u\|_1 + c_2 \|f_u - f_v\|_1$$
$$\leq c_1 |p(u) - p(v)| \|f_0\|_1 + c_2 \|f_u - f_v\|_1$$

as required.

Lemma 2.2.2. *Under the above conditions we have*

$$(d/ds) \|f_s\|_{p(s)}^{p(s)} = p(s) \langle f_s', f_s^{p(s)-1} \rangle + p'(s) \int_\Omega f_s^{p(s)} \log f_s \, dx.$$

Proof. We have to show that $f_s^{p(s)}$ is a continuously differentiable function of s with values in L^1, and that

$$(d/ds) f_s^{p(s)} = p(s) f_s' f_s^{p(s)-1} + p'(s) f_s^{p(s)} \log f_s.$$

This is equivalent to

$$f_s^{p(s)} - f_0^{p(0)} = \int_0^s \{p(u) f_u' f_u^{p(u)-1} + p'(u) f_u^{p(u)} \log f_u\} \, du \quad (2.2.1)$$

both sides taking values in L^1, the integrand being a norm continuous function of u by Lemma 2.2.1. The validity of this formula is proved by approximating f_s by continuous piecewise linear functions of s, for which (2.2.1) holds at each point $x \in \Omega$ by elementary calculus.

Theorem 2.2.3. *Let e^{-Ht} be ultracontractive with*

$$\|e^{-Ht}\|_{\infty, 2} \leq e^{M(t)}$$

for all $t > 0$, where $M(t)$ is a monotonically decreasing continuous function of

t. Then $0 \leq f \in \text{Quad}(H) \cap L^1 \cap L^\infty$ implies $f^2 \log f \in L^1$, and the logarithmic Sobolev inequality

$$\int_\Omega f^2 \log f \, dx \leq \varepsilon Q(f) + M(\varepsilon) \|f\|_2^2 + \|f\|_2^2 \log \|f\|_2 \qquad (2.2.2)$$

is valid for all $\varepsilon > 0$.

Proof. We start by assuming that $\|f\|_2 = 1$ and $0 \leq f \in \text{Dom}(H) \cap L^1 \cap L^\infty$. By the Stein interpolation theorem we have

$$\|e^{-Hs}\|_{p(s),2} \leq \exp\{M(t)s/t\}$$

for all $0 \leq s < t$, where $p(s) = 2t/(t-s)$. Therefore

$$\|f_s\|_{p(s)}^{p(s)} \leq \exp\{M(t)sp(s)/t\}$$

and

$$(d/ds)\|f_s\|_{p(s)}^{p(s)}|_{s=0} \leq 2M(t)/t.$$

Putting $s = 0$ in Lemma 2.2.2 we deduce that

$$-t\langle Hf, f\rangle + \int_\Omega f^2 \log f \, dx \leq M(t)$$

and putting $\varepsilon = t$ we obtain

$$\int_\Omega f^2 \log f \, dx \leq \varepsilon Q(f) + M(\varepsilon). \qquad (2.2.3)$$

If $0 \leq g \in \text{Dom}(H) \cap L^1 \cap L^\infty$ then putting $f = g/\|g\|_2$ in (2.2.3) yields (2.2.2), with g in place of f.

Now let $0 \leq g \in \text{Quad}(H) \cap L^1 \cap L^\infty$ and put $g_\delta = e^{-H\delta} g$ where $\delta > 0$, so that $g_\delta \in \text{Dom}(H) \cap L^1 \cap L^\infty$ and

$$\int g_\delta^2 \log g_\delta \, dx \leq \varepsilon Q(g_\delta) + M(\varepsilon) \|g_\delta\|_2^2 + \|g_\delta\|_2^2 \log \|g_\delta\|_2.$$

It is elementary that $Q(g_\delta) \to Q(g)$ and $\|g_\delta\|_2 \to \|g\|_2$ as $\delta \to 0$, so we need only examine the limit of the LHS. Since

$$g_\delta(x) + g_\delta^2(x) \log g_\delta(x) \geq 0$$

for all $\delta > 0$ and $x \in \Omega$, Fatou's lemma yields

$$\int_\Omega (g + g^2 \log g) \, dx \leq \liminf_{\delta \to 0} \int_\Omega (g_\delta + g_\delta^2 \log g_\delta) \, dx.$$

But

$$\int_\Omega g \, dx = \lim_{\delta \to 0} \int g_\delta \, dx$$

so

$$\int g^2 \log g \, dx \leq \liminf_{\delta \to 0} \int g_\delta^2 \log g_\delta \, dx$$

as required to complete the proof.

If we only assume that $0 \leq f \in \text{Quad}(H)$ then it is not obvious that $f^2 \log f \in L^1$, so we have a slight variation of the above theorem. The constant $\frac{1}{4}$ in the formula for $\beta(\varepsilon)$ below is clearly not optimal.

Theorem 2.2.4. Let e^{-Ht} be ultracontractive with

$$\|e^{-Ht}\|_{\infty,2} \leq e^{M(t)} \qquad (2.2.4)$$

for all $t > 0$, where $M(t)$ is a monotonically decreasing continuous function of t. Then $0 \leq f \in \text{Quad}(H)$ implies

$$\int_\Omega f^2 \log_+ f \, dx \leq \varepsilon Q(f) + \bar{\beta}(\varepsilon)\|f\|_2^2 + \|f\|_2^2 \log \|f\|_2 \qquad (2.2.5)$$

for all $\varepsilon > 0$ where

$$\bar{\beta}(\varepsilon) = M(\varepsilon/4) + 2.$$

Proof. We first suppose that $0 \leq f \in \text{Quad}(H) \cap L^\infty$. If we put

$$g(x) = \begin{cases} f(x) & \text{if } f(x) \geq 1 \\ 2f(x) - 1 & \text{if } \frac{1}{2} \leq f(x) < 1 \\ 0 & \text{if } 0 \leq f(x) < \frac{1}{2} \end{cases}$$

and

$$E = \{x : f(x) \geq \tfrac{1}{2}\}$$

then $0 \leq g \in \text{Quad}(H) \cap L^1 \cap L^\infty$ and

$$0 \leq g = g\chi_E \leq f.$$

Moreover

$$0 \leq \int_\Omega g^2 \log_- g \, dx \leq \|g\|_1 \leq \|g\|_2 \|\chi_E\|_2 \leq 2\|f\|_2^2$$

and

$$Q(g) \leq 4Q(f)$$

by Theorem 1.3.3 Condition (iii). Therefore

$$\int_\Omega f^2 \log_+ f \, dx = \int_\Omega g^2 \log_+ g \, dx$$

$$\leq \int_\Omega g^2 \log g \, dx + 2\|f\|_2^2$$

$$\leq \varepsilon Q(g) + M(\varepsilon)\|g\|_2^2 \log \|g\|_2 + 2\|f\|_2^2$$

$$\leq 4\varepsilon Q(f) + (M(\varepsilon) + 2)\|f\|_2^2 + \|f\|_2^2 \log \|f\|_2.$$

The validity of the theorem for such f now follows.

The fundamental inequalities

Finally given $0 \leq f \in \text{Quad}(H)$, we put $f_n = f \wedge n$ so that $\|f_n\|_2 \to \|f\|_2$ as $n \to \infty$ and $Q(f_n) \leq Q(f)$ for all n by Theorem 1.3.3 Condition (iii). The validity of the theorem now follows from the monotone convergence theorem.

Note 2.2.5. If one only assumes that e^{-Ht} is hypercontractive, then one obtains a bound similar to (2.2.5) for *some* positive constants ε, β. We leave the reader to work out the details.

We now turn to the converse of the above theorems.

Lemma 2.2.6. *Suppose that there exists a monotonically decreasing continuous function $\beta(\varepsilon)$ such that*

$$\int_\Omega f^2 \log f \, dx \leq \varepsilon Q(f) + \beta(\varepsilon) \|f\|_2^2 + \|f\|_2^2 \log \|f\|_2 \qquad (2.2.6)$$

for all $\varepsilon > 0$ and $0 \leq f \in \text{Quad}(H) \cap L^1 \cap L^\infty$. Then

$$\int g^p \log g \, dx \leq \varepsilon \langle Hg, g^{p-1} \rangle + 2\beta(\varepsilon) p^{-1} \|g\|_p^p + \|g\|_p^p \log \|g\|_p$$

for all $2 < p < \infty$ and all $g \in \mathcal{D}_+ \equiv \bigcup_{t>0} e^{-Ht}(L^1 \cap L^\infty)_+$.

Proof. Putting $f = g^{p/2}$ in (2.2.6) we obtain

$$(p/2) \int g^p \log g \, dx \leq \varepsilon Q(g^{p/2}) + \beta(\varepsilon) \|g\|_p^p + \frac{p}{2} \|g\|_p^p \log \|g\|_p$$

so subject to the inequality

$$Q(g^{p/2}) \leq \frac{p^2}{4(p-1)} \langle Hg, g^{p-1} \rangle \qquad (2.2.7)$$

we obtain

$$\int g^p \log g \, dx \leq \frac{\varepsilon p}{2(p-1)} \langle Hg, g^{p-1} \rangle + \frac{2\beta(\varepsilon)}{p} \|g\|_p^p + \|g\|_p^p \log \|g\|_p$$

which yields the result.

The proof of (2.2.7) depends upon the formulae

$$Q(g^{p/2}) = \lim_{t \to 0} t^{-1} \langle (1 - e^{-Ht}) g^{p/2}, g^{p/2} \rangle,$$

$$\langle Hg, g^{p-1} \rangle = \lim_{t \to 0} t^{-1} \langle (1 - e^{-Ht}) g, g^{p-1} \rangle.$$

It is clearly sufficient to prove that

$$\langle (1 - e^{-Ht}) g^{p/2}, g^{p/2} \rangle \leq \frac{p^2}{4(p-1)} \langle (1 - e^{-Ht}) g, g^{p-1} \rangle \qquad (2.2.8)$$

for all $0 < t < \infty$, $2 < p < \infty$ and $0 \leq g \in L^1 \cap L^\infty$.

We recall from the proof of Theorem 1.3.3 that there exist a symmetric Borel measure $\mu_t \geq 0$ on $\Omega \times \Omega$ and a function $\rho_t: \Omega \to [0,1]$ such that

$$\langle e^{-Ht} u, v \rangle = \int_{\Omega \times \Omega} u(x)\overline{v(y)}\, d\mu_t(x,y)$$

and

$$\int_{\Omega \times \Omega} u(x)\, d\mu_t(x,y) = \int_\Omega \rho_t(x) u(x)\, dx$$

for all $u, v \in C_c(\Omega)$, and hence for all $u, v \in L^1 \cap L^\infty$. It follows that

$$\langle (1 - e^{-Ht})g^{p/2}, g^{p/2} \rangle = \int_\Omega \{1 - \rho_t(x)\} g(x)^p\, dx$$

$$+ \tfrac{1}{2} \int_{\Omega \times \Omega} |g(x)^{p/2} - g(y)^{p/2}|^2\, d\mu_t(x,y)$$

and

$$\langle (1 - e^{-Ht})g, g^{p-1} \rangle = \int_\Omega \{1 - \rho_t(x)\} g(x)^p\, dx$$

$$+ \tfrac{1}{2} \int_{\Omega \times \Omega} \{g(x) - g(y)\}\{g(x)^{p-1} - g(y)^{p-1}\}\, d\mu_t(x,y).$$

Since $p^2/4(p-1) \geq 1$ for all $2 < p < \infty$, (2.2.8) follows provided we can establish that

$$\int_{\Omega \times \Omega} |g(x)^{p/2} - g(y)^{p/2}|^2\, d\mu_t(x,y) \leq \frac{p^2}{4(p-1)} \int_{\Omega \times \Omega} \{g(x) - g(y)\}$$

$$\cdot \{g(x)^{p-1} - g(y)^{p-1}\}\, d\mu_t(x,y).$$

This will follow from the inequality

$$|\alpha^{p/2} - \beta^{p/2}|^2 \leq \frac{p^2}{4(p-1)} (\alpha - \beta)(\alpha^{p-1} - \beta^{p-1}) \qquad (2.2.9)$$

valid for all $\alpha \geq 0$, $\beta \geq 0$ and $2 < p < \infty$. By symmetry we need only prove (2.2.9) for the case $0 \leq \beta \leq \alpha < \infty$. Using the Schwartz inequality we have

$$|\alpha^{p/2} - \beta^{p/2}|^2 = \left| \int_\beta^\alpha \frac{p}{2} s^{p/2 - 1}\, ds \right|^2$$

$$\leq \frac{p^2}{4}(\alpha - \beta) \int_\beta^\alpha s^{p-2}\, ds$$

$$= \frac{p^2}{4(p-1)}(\alpha - \beta)(\alpha^{p-1} - \beta^{p-1}).$$

Note. If H is an elliptic differential operator then the explicit expression for the quadratic form shows that (2.2.7) is an equality, so most of the above computations become unnecessary.

Theorem 2.2.7. *Let $\varepsilon(p) > 0$ and $\Gamma(p)$ be two continuous functions defined for $2 < p < \infty$ such that*

$$\int_\Omega f^p \log f \, dx \leqslant \varepsilon(p) \langle Hf, f^{p-1} \rangle + \Gamma(p) \|f\|_p^p + \|f\|_p^p \log \|f\|_p$$

for all $2 < p < \infty$ and $f \in \mathscr{D}_+ = \bigcup_{t>0} e^{-Ht}(L^1 \cap L^\infty)_+$. If

$$t = \int_2^\infty p^{-1} \varepsilon(p) \, dp, \qquad M = \int_2^\infty p^{-1} \Gamma(p) \, dp$$

are both finite then e^{-Ht} maps L^2 into L^∞ and

$$\|e^{-Ht}\|_{\infty, 2} \leqslant e^M.$$

Proof. We define the function $p(s)$ for $0 \leqslant s < t$ by

$$dp/ds = p/\varepsilon(p), \qquad p(0) = 2 \qquad (2.2.10)$$

so that $p(s)$ is monotonically increasing and $p(s) \to \infty$ as $s \to t$. We define the function $N(s)$ for $0 \leqslant s < t$ by

$$dN/ds = \Gamma(p)/\varepsilon(p), \qquad N(0) = 0$$

so that $N(s) \to M$ as $s \to t$. If $f \in \mathscr{D}_+$ and $f_s = e^{-Hs}f$ for $0 < s < t$ then

$$\frac{d}{ds} \log \{e^{-N(s)} \|f_s\|_{p(s)}\} = \frac{d}{ds} \left\{ -N(s) + \frac{1}{p(s)} \log \|f_s\|_{p(s)}^{p(s)} \right\}$$

$$= -\frac{\Gamma}{\varepsilon} - \frac{1}{p^2} \frac{p}{\varepsilon} \log \|f_s\|_p^p + \frac{1}{p} \|f_s\|_p^{-p} \left\{ -p \langle Hf_s, f_s^{p-1} \rangle \right.$$

$$\left. + \frac{p}{\varepsilon} \int_\Omega f_s^p \log f_s \, dx \right\} = \varepsilon^{-1} \|f_s\|_p^{-p} \left\{ \int_\Omega f_s^p \log f_s \, dx \right.$$

$$\left. - \varepsilon \langle Hf_s, f_s^{p-1} \rangle - \Gamma \|f_s\|_p^p - \|f_s\|_p^p \log \|f_s\|_p \right\}$$

$$\leqslant 0.$$

Therefore

$$e^{-N(s)} \|f_s\|_{p(s)} \leqslant \|f\|_2$$

for all $0 \leqslant s < t$. It follows that

$$\|e^{-Ht}f\|_{p(s)} \leqslant \|e^{-Hs}f\|_{p(s)} \leqslant e^{N(s)} \|f\|_2$$

for all $0 \leqslant s < t$, whence

$$\|e^{-Ht}f\|_\infty \leqslant e^M \|f\|_2.$$

Finally if $0 \leqslant f \in L^2$ there exists a sequence of $f_n \in \mathscr{D}_+$ such that

$\|f_n - f\|_2 \to 0$. Since $\|e^{-Ht}f_n - e^{-Ht}f\|_2 \to 0$ and
$$\|e^{-Ht}f_n\|_\infty \leq e^M \|f_n\|_2$$
by the above calculations, we deduce that
$$\|e^{-Ht}f\|_\infty \leq e^M \|f\|_2.$$
For a general $f \in L^2$ we have
$$|e^{-Ht}f| \leq e^{-Ht}|f|$$
by the positivity of e^{-Ht}, so
$$\|e^{-Ht}f\|_\infty \leq \|e^{-Ht}|f|\|_\infty$$
$$\leq e^M \||f|\|_2$$
$$= e^M \|f\|_2$$
which concludes the proof.

We shall often apply the above theorem by choosing
$$\varepsilon(p) = 2t/p, \quad \Gamma(p) = 2\beta(\varepsilon(p))/p. \tag{2.2.11}$$
Other choices of $\varepsilon(p)$ are sometimes necessary to deal with problems for which $\beta(\varepsilon)$ diverges rapidly as $\varepsilon \to 0$, and one can even choose $\varepsilon(p)$ so as to minimise M for fixed $t > 0$ in Theorem 2.2.7. If we make the choice (2.2.11) then the solution of (2.2.10) is
$$p(s) = 2t/(t-s). \tag{2.2.12}$$
Note that this is the same function as in the proof of Theorem 2.2.3.

Corollary 2.2.8. *Let $\beta(\varepsilon)$ be a monotonically decreasing continuous function of ε such that*
$$\int_\Omega f^2 \log f \, dx \leq \varepsilon Q(f) + \beta(\varepsilon)\|f\|_2^2 + \|f\|_2^2 \log \|f\|_2$$
for all $\varepsilon > 0$ and $0 \leq f \in \text{Quad}(H) \cap L^1 \cap L^\infty$. Suppose that
$$M(t) = (1/t) \int_0^t \beta(\varepsilon) \, d\varepsilon$$
is finite for all $t > 0$. Then e^{-Ht} is ultracontractive and
$$\|e^{-Ht}\|_{\infty,2} \leq e^{M(t)}$$
for all $0 < t < \infty$.

Proof. Applying Lemma 2.2.6 and Theorem 2.2.7 with the choices (2.2.11) and (2.2.12) we obtain the required result with
$$M(t) = \int_2^\infty 2\beta(\varepsilon(p))p^{-2} \, dp$$
$$= (1/t) \int_0^t \beta(\varepsilon) \, d\varepsilon.$$

Note 2.2.9. A stronger form of the hypothesis of Corollary 2.2.8 is that

$$\int_\Omega f^2 \log_+ f \, dx \leq \varepsilon Q(f) + \beta(\varepsilon)\|f\|_2^2 + \|f\|_2^2 \log \|f\|_2 \quad (2.2.13)$$

for all $\varepsilon > 0$ and $0 \leq f \in \mathrm{Quad}\,(H)$. By Fatou's lemma we need only assume (2.2.13) for a subset of $\{f \geq 0 : f \in \mathrm{Quad}\,(H)\}$ which is dense for the norm

$$\|\|f\|\| = \{\|f\|^2 + Q(f)\}^{\frac{1}{2}}.$$

2.3. Examples and applications

In this section we examine a series of special cases of the above theory to see in what sense Theorems 2.2.3 and 2.2.7 are converses of each other. Some of these examples will be of fundamental significance for later work. In all cases we examine the relationship between the bounds

$$\|e^{-Ht}\|_{\infty,2} \leq e^{M(t)} \quad (2.3.1)$$

for all $0 < t < \infty$, and

$$\int_\Omega f^2 \log f \, dx \leq \varepsilon Q(f) + \beta(\varepsilon)\|f\|_2^2 + \|f\|_2^2 \log \|f\|_2 \quad (2.3.2)$$

for all $0 < \varepsilon < \infty$ and $0 \leq f \in \mathrm{Quad}\,(H) \cap L^1 \cap L^\infty$.

In all the examples below explicit expressions for the constants c_i may be obtained by examining the details of the proofs. One does not normally, however, obtain the minimum values of these constants by this method. We do not assume that N is the dimension of Ω below, nor even that N or P are integers.

Example 2.3.1. If there exist constants $c_1 > 0$ and $N \geq 0$ such that

$$e^{M(t)} \leq c_1 t^{-N/4}$$

for all $t > 0$, then there exists a constant $c_2 > 0$ such that

$$\beta(\varepsilon) \leq c_2 - (N/4)\log \varepsilon \quad (2.3.3)$$

for all $\varepsilon > 0$. Conversely (2.3.3) implies that there exists a constant $c_3 > 0$ such that

$$e^{M(t)} \leq c_3 t^{-N/4}$$

for all $t > 0$; the proof uses Corollary 2.2.8.

Example 2.3.2. If there exist constants $c_1 > 0, N \geq 0$ and $p \geq 0$ such that

$$e^{M(t)} \leq \begin{cases} c_1 t^{-N/4} & \text{if } 0 < t \leq 1 \\ c_1 t^{-P/4} & \text{if } 1 \leq t < \infty \end{cases}$$

then there exists a constant $c_2 > 0$ such that

$$\beta(\varepsilon) \leqslant \begin{cases} c_2 - (N/4)\log\varepsilon & \text{if } 0 < \varepsilon \leqslant 1 \\ c_2 - (P/4)\log\varepsilon & \text{if } 1 \leqslant \varepsilon < \infty. \end{cases} \qquad (2.3.4)$$

Conversely (2.3.4) implies that there exists a constant $c_3 > 0$ such that

$$e^{M(t)} \leqslant \begin{cases} c_3 t^{-N/4} & \text{if } 0 < t \leqslant 1 \\ c_3 t^{-P/4} & \text{if } 1 \leqslant t < \infty. \end{cases}$$

The proof relies upon Corollary 2.2.8.

Example 2.3.3. If there exist constants $c_1 > 0$, $N \geqslant 0$ and $E \geqslant 0$ such that

$$e^{M(t)} \leqslant \begin{cases} c_1 t^{-N/4} & \text{if } 0 < t \leqslant 1 \\ c_1 e^{-E(t-1)} & \text{if } 1 \leqslant t < \infty \end{cases}$$

then there exists a constant $c_2 > 0$ such that

$$\beta(\varepsilon) \leqslant \begin{cases} c_2 - (N/4)\log\varepsilon & \text{if } 0 < \varepsilon \leqslant 1 \\ c_2 - E(\varepsilon - 1) & \text{if } 1 \leqslant \varepsilon < \infty. \end{cases} \qquad (2.3.5)$$

Conversely from (2.3.3) and the case $P = 0$ of Example 2.3.2 we obtain

$$e^{M(t)} \leqslant c_3 t^{-N/4} \quad \text{if } 0 < t \leqslant 1.$$

Also substituting (2.3.5) into (2.3.2) and letting $\varepsilon \to \infty$ we find that

$$E\|f\|^2 \leqslant Q(f) \qquad (2.3.6)$$

for all $0 \leqslant f \in \text{Quad}(H) \cap L^1 \cap L^\infty$. Since $Q(|f|) \leqslant Q(f)$ for all f we deduce that (2.3.6) also holds for all $f \in \text{Quad}(H) \cap L^1 \cap L^\infty$, and hence for all f in

$$\mathcal{D} = (H+1)^{-1}(L^1 \cap L^\infty).$$

But $\mathcal{D}$ is evidently an operator core of H so $H \geqslant E$. If $t \geqslant 1$ we now can conclude that

$$\|e^{-Ht}f\|_\infty \leqslant c_3 \|e^{-H(t-1)}f\|_2$$
$$\leqslant c_3 e^{-E(t-1)}\|f\|_2$$

so

$$e^{M(t)} \leqslant c_3 e^{-E(t-1)}$$

for all $1 \leqslant t < \infty$.

Example 2.3.4. If there exist constants $c_1 > 0$ and $\alpha \in \mathbb{R}$ such that

$$M(t) \leqslant c_1(1 + t^{-\alpha})$$

for all $t > 0$ then

$$\beta(\varepsilon) \leqslant c_1(1 + \varepsilon^{-\alpha}) \qquad (2.3.7)$$

for all $\varepsilon > 0$. The converse of this result can only be obtained from Corollary 2.2.8 when $0 < \alpha < 1$. For $\alpha \geqslant 1$ we have to go back to

Examples and applications

Theorem 2.2.7. If we put

$$\varepsilon(p) = \gamma t 2^\gamma p^{-\gamma}$$

where $2\gamma\alpha = 1$, then a direct computation verifies that

$$t = \int_2^\infty p^{-1}\varepsilon(p)\,dp.$$

Putting

$$\Gamma(p) = 2\beta(\varepsilon(p))p^{-1}$$
$$\leqslant 2c_1 p^{-1} + c_2 t^{-\alpha} p^{-1+\gamma\alpha}$$
$$= 2c_1 p^{-1} + c_2 t^{-\alpha} p^{-\frac{1}{2}}$$

in Theorem 2.2.2, we obtain

$$M(t) \leqslant c_3(1 + t^{-\alpha}).$$

Example 2.3.5. As a final example we record the fact that if

$$\beta(\varepsilon) \leqslant c_1 \exp(\varepsilon^{-\alpha})$$

for all $0 < \varepsilon < \infty$ and some $0 < \alpha < 1$, then e^{-Ht} is ultracontractive. However if $\alpha = 1$, then this need not be true.

Typical of the applications of the above abstract results is the following.

Theorem 2.3.6. Let H be a strictly elliptic operator on $L^2(\Omega)$ where Ω is a region in $\mathbb{R}^N$. Then e^{-Ht} has a kernel $K(t, x, y)$ such that

$$0 \leqslant K(t, x, y) \leqslant ct^{-N/2}$$

almost everywhere on $\Omega \times \Omega$ for all $t > 0$.

Proof. Let $H_0 = -\Delta$ on $L^2(\Omega)$ with the usual Dirichlet boundary conditions. By Section 1.1.8 and Lemma 2.2.6 we see that the kernel $K_0(t, x, y)$ of $e^{-H_0 t}$ satisfies

$$0 \leqslant K_0(t, x, y) \leqslant (4\pi t)^{-N/2} e^{-(x-y)^2/4t}$$
$$\leqslant (4\pi t)^{-N/2}$$

for all $t > 0$. It follows that

$$\|e^{-H_0 t} f\|_\infty \leqslant c_1 t^{-N/4} \|f\|_2$$

for all $f \in L^2(\Omega)$ and $t > 0$, and hence by Example 2.3.1 that

$$\int_\Omega f^2 \log f\,dx \leqslant \varepsilon Q_0(f) + \beta_0(\varepsilon)\|f\|_2^2 + \|f\|_2^2 \log \|f\|_2 \quad (2.3.8)$$

for all $0 < \varepsilon < \infty$ and $0 \leqslant f \in \mathrm{Quad}(H_0) \cap L^1 \cap L^\infty$, where

$$\beta_0(\varepsilon) = c_2 - (N/4)\log\varepsilon.$$

But H is strictly elliptic so
$$Q_0(f) \leq c_3 Q(f)$$
for all f and (2.3.8) implies that
$$\int_\Omega f^2 \log f \, dx \leq \varepsilon Q(f) + \beta(\varepsilon)\|f\|_2^2 + \|f\|_2^2 \log \|f\|_2$$
for all $0 < \varepsilon < \infty$ and $0 \leq f \in \text{Quad}(H) \cap L^1 \cap L^\infty$, where
$$\beta(\varepsilon) = c_4 - (N/4) \log \varepsilon.$$
The proof is completed by applying Example 2.3.1 in the reverse direction.

As a final application of the ideas of Section 2.3 we consider a variation of Example 2.3.3 in which we weaken the requirement that e^{-Ht} is a symmetric Markov semigroup.

Example 2.3.7. Let e^{-Ht} be a self-adjoint positivity-preserving semigroup on $L^2(\Omega, dx)$ such that $\text{Sp}(H) \subseteq [E, \infty)$ and
$$\|e^{-Ht} f\|_\infty \leq e^{Ft} \|f\|_\infty \quad (2.3.9)$$
for all $f \in L^\infty$ and $t \geq 0$. If also (2.3.2) holds for $0 < \varepsilon \leq 1$ and for
$$\beta(\varepsilon) = c_0 - (N/4) \log \varepsilon$$
then
$$\|e^{-Ht} f\|_\infty \leq c(t) \|f\|_1$$
for all $f \in L^1$ and $t > 0$, where
$$c(t) = \begin{cases} c_1 t^{-N/2} & \text{if } 0 < t \leq 1 \\ c_1 e^{-E(t-1)} & \text{if } 1 \leq t < \infty. \end{cases}$$

Proof. Since $e^{-(H+F)t}$ is a symmetric Markov semigroup we may conclude as in Example 2.3.3 that
$$\|e^{-(H+F)t} f\|_\infty \leq c_2 t^{-N/4} \|f\|_2$$
for all $f \in L^2$ and $0 < t \leq 1$.

Therefore if $0 < t \leq 1$ we have
$$\|e^{-Ht} f\|_\infty \leq c_2 e^{Ft/2} (t/2)^{-N/4} \|e^{-Ht/2} f\|_2$$
$$\leq c_1 t^{-N/2} \|f\|_1$$
while if $1 \leq t < \infty$
$$\|e^{-Ht} f\|_\infty \leq \|e^{-H/2}\|_{\infty,2} \|e^{-H(t-1)}\|_{2,2} \|e^{-H/2}\|_{2,1} \|f\|_1$$
$$\leq c_2 e^{-E(t-1)} \|f\|_1.$$

Note. According to the above method of calculation c_1 depends upon c_0, N and F. We do not know whether the hypothesis (2.3.9) is necessary for the result.

2.4 Ultracontractivity, Sobolev and Nash inequalities

Although the deepest insight into ultracontractivity uses logarithmic Sobolev inequalities, a number of the commonest applications can be resolved using ordinary Sobolev inequalities. We present this alternative approach here. Our first result is easy, but not quite sharp. As before we take e^{-Ht} to be a symmetric Markov semigroup on $L^2(\Omega)$.

Theorem 2.4.1. *If $\mu > 0$ and*
$$\|f\|_\infty \leqslant c_1 \|(H+1)^\mu f\|_2 \tag{2.4.1}$$
for all $f \in \mathrm{Dom}(H+1)^\mu$, then
$$\|e^{-Ht}f\|_\infty \leqslant c_2 \max(1, t^{-\mu}) \|f\|_2 \tag{2.4.2}$$
for all $f \in L^2(\Omega)$ and $t > 0$. Conversely (2.4.2) implies
$$\|f\|_\infty \leqslant c_{3,\nu} \|(H+1)^\nu f\|_2$$
for all $\nu > \mu$ and $f \in \mathrm{Dom}(H+1)^\nu$.

Proof. For the first part it is sufficient to prove (2.4.2) where $0 < t < 1$. This follows from the facts that (2.4.1) is equivalent to
$$\|(H+1)^{-\mu} f\|_\infty \leqslant c_1 \|f\|_2$$
for all $f \in L^2(\Omega)$ and that
$$\|(H+1)^\mu e^{-Ht} f\|_2 \leqslant c_4 t^{-\mu} \|f\|_2$$
for all $f \in L^2(\Omega)$ and $0 < t < 1$.

Conversely given (2.4.2) we deduce (2.4.1) by use of the formula
$$(H+1)^{-\nu} f = \Gamma(\nu)^{-1} \int_0^\infty t^{\nu-1} e^{-(H+1)t} f \, dt$$
which converges in L^∞ norm if $\nu > \mu$.

We give two slightly different versions of our next result, which is much deeper.

Theorem 2.4.2. *If $\mu > 2$ then a bound of the form*
$$\|e^{-Ht} f\|_\infty \leqslant c_1 t^{-\mu/4} \|f\|_2 \tag{2.4.3}$$
for all $t > 0$ and all $f \in L^2(\Omega)$, is equivalent to a bound of the form
$$\|f\|_{2\mu/(\mu-2)}^2 \leqslant c_2 Q(f) \tag{2.4.4}$$
for all $f \in \mathrm{Quad}(H)$.

Proof. We combine the fact that e^{-Ht} is a contraction on L^∞ with the bound
$$\|e^{-Ht} f\|_\infty \leqslant c t^{-\mu/2} \|f\|_1$$

deduced from (2.4.3). The Riesz–Thorin interpolation theorem yields
$$\|e^{-Ht}f\|_\infty \leq ct^{-\mu/2q}\|f\|_q$$
for all $t > 0$ and $f \in L^q$, where we take q to satisfy $1 < q < \mu$. We now write
$$H^{-\frac{1}{2}}f = g + h$$
where
$$g = \Gamma(\tfrac{1}{2})^{-1}\int_0^T t^{-\frac{1}{2}}e^{-Ht}f\,dt$$
$$h = \Gamma(\tfrac{1}{2})^{-1}\int_T^\infty t^{-\frac{1}{2}}e^{-Ht}f\,dt.$$

We see that
$$\|h\|_\infty = \Gamma(\tfrac{1}{2})^{-1}\int_T^\infty ct^{-\frac{1}{2}-\mu/2q}\|f\|_q\,dt$$
$$= c\|f\|_q T^{\frac{1}{2}-\mu/2q}.$$

Given $\lambda > 0$ we define T by
$$\lambda/2 = c\|f\|_q T^{\frac{1}{2}-\mu/2q}$$
so that
$$|\{x:|H^{-\frac{1}{2}}f(x)|\geq\lambda\}| \leq |\{x:|g(x)|\geq\lambda/2\}| \leq (\lambda/2)^{-q}\|g\|_q^q$$
$$\leq (\lambda/2)^{-q}\{\Gamma(\tfrac{1}{2})^{-1}2T^{\frac{1}{2}}\|f\|_q\}^q$$
since e^{-Ht} is a contraction on L^q. Putting
$$1/r = 1/q - 1/\mu$$
we deduce that
$$|\{x:|H^{-\frac{1}{2}}f(x)|\geq\lambda\}| \leq c\lambda^{-q}(\lambda/\|f\|_q)^{q/(1-\mu/q)}\|f\|_q^q$$
$$= c\lambda^{-r}\|f\|_q^r.$$

This shows that $H^{-\frac{1}{2}}$ is of weak type (q,r) for all $1 < q < \mu$. By applying the Marcinkiewicz interpolation theorem we deduce that $H^{-\frac{1}{2}}$ is bounded from L^2 to L^p where
$$1/p = \tfrac{1}{2} - 1/\mu \tag{2.4.5}$$
and this is equivalent to the bound (2.4.4).

For the converse we assume (2.4.4) and define p by (2.4.5). If $f \geq 0$ satisfies $\|f\|_2 = 1$, then we define the measure μ of total mass 1 by
$$d\mu(x) = f(x)^2\,dx.$$

Since log is concave we have
$$\int f^2 \log f\,dx = \frac{1}{p-2}\int \log(f^{p-2})\,d\mu$$

$$\leq \frac{1}{p-2}\log \int f^{p-2}\, d\mu$$

$$= \frac{1}{p-2}\log \|f\|_p^p$$

$$\leq \frac{p}{2(p-2)}\log \|f\|_p^2$$

$$\leq (\mu/4)(-\log \varepsilon + \varepsilon \|f\|_p^2)$$

$$\leq (\mu/4)\{-\log \varepsilon + c_2\varepsilon Q(f)\}.$$

By redefining ε we obtain the hypothesis (2.3.3) of Example 2.3.1.

Corollary 2.4.3. *If $\mu > 2$ then a bound of the form*

$$\|e^{-Ht}f\|_\infty \leq c_1 t^{-\mu/4}\|f\|_2$$

for all $0 < t < 1$ and all $f \in L^2(\Omega)$ is equivalent to a bound of the form

$$\|f\|_{2\mu/(\mu-2)}^2 \leq c_2\{Q(f) + \|f\|_2^2\}$$

for all $f \in \text{Quad}(H)$.

Proof. The above hypotheses imply the hypotheses of Theorem 2.4.2 for the operator $(H+1)$, which clearly still generates a symmetric Markov semigroup.

Although the above theorems are restricted to the case $\mu > 2$, this can often be circumvented in applications, among which the following are typical.

Theorem 2.4.4. *Let H be a uniformly elliptic operator with Neumann boundary conditions acting on $L^2(\Omega)$, where Ω is a region in $\mathbb{R}^N$ with the extension property. Then e^{-Ht} has a kernel $K(t,x,y)$ which satisfies*

$$0 \leq K(t,x,y) \leq ct^{-N/2}$$

almost everywhere on $\Omega \times \Omega$ for all $0 < t < 1$.

Proof. First assume that $N \geq 3$. Then we have

$$\|f\|_{2N/(N-2)} \leq c_1(\|\nabla f\|_2^2 + \|f\|_2^2)$$
$$\leq c_2\{Q(f) + \|f\|_2^2\}$$

by Lemma 1.7.11, and can apply Corollary 2.4.3.

If $N \leq 2$ then we consider instead the operator

$$H' = H \otimes 1 + 1 \otimes (-\Delta)$$

on $L^2(\Omega \times \mathbb{R}^2)$ and use the above method to obtain the bound

$$0 \leq K(t,x,y)K'(t,x',y') \leq ct^{-(N+2)/2}$$

for $0 < t < 1$ where
$$K'(t, x', y') = (4\pi t)^{-1} \exp\{-(x'-y')^2/4t\}.$$
The result now follows upon putting $x' = y'$.

The following variant of Theorem 2.4.2 is often useful. There is also a similar version of Corollary 2.4.3.

Theorem 2.4.5. *If $\mu > 2$ then the Condition (2.4.4) of Theorem 2.4.2 is equivalent to*
$$g \leqslant c_2 \|g\|_{\mu/2} H \qquad (2.4.6)$$
for all $g \in L^{\mu/2}(\Omega)$.

Proof. If (2.4.5) holds and $f \in \text{Quad } H$ then
$$\langle gf, f \rangle \leqslant \|g\|_{\mu/2} \|f^2\|_v = \|g\|_{\mu/2} \|f\|_{2v}^2$$
where
$$2/\mu + 1/v = 1.$$
Therefore
$$\langle gf, f \rangle \leqslant \|g\|_{\mu/2} \|f\|_{2\mu/(\mu-2)}$$
$$\leqslant c_2 \|g\|_{\mu/2} Q(f)$$
which implies (2.4.6). Conversely given (2.4.6) we have
$$\int g|f|^2 \leqslant c_2 \|g\|_{\mu/2} Q(f)$$
for all $f \in \text{Quad}(H)$. Putting $g = |f|^{4/(\mu-2)}$ yields
$$\int |f|^{2\mu/(\mu-2)} \leqslant c_2 \left(\int f^{2\mu/(\mu-2)} \right)^{2/\mu} Q(f)$$
which implies (2.4.4).

There is yet another approach to ultracontractive estimates which uses inequalities associated with Nash, and which is even more direct. Unlike Theorem 2.4.2 it does not require that $\mu > 2$.

Theorem 2.4.6. *Let e^{-Ht} be a symmetric Markov semigroup on $L^2(\Omega)$ and let $\mu > 0$. Then the following two bounds are equivalent:*

(i)
$$\|e^{-Ht} f\|_\infty \leqslant c_1 t^{-\mu/4} \|f\|_2$$
for some $c_1 < \infty$, all $t > 0$ and all $f \in L^2$.

(ii)
$$\|f\|_2^{2+4/\mu} \leqslant c_2 Q(f) \|f\|_1^{4/\mu}$$
for some $c_2 < \infty$ and all $0 \leqslant f \in \text{Dom}(Q) \cap L^1$.

Proof. Given (i) we have by duality

$$c_1^2 t^{-\mu/2} \|f\|_1^2 \geq \langle e^{-H2t}f, f \rangle$$
$$= \langle f, f \rangle - \int_0^{2t} Q(e^{-Hs/2}f)\,ds$$
$$\geq \langle f, f \rangle - 2tQ(f)$$

for all $0 \leq f \in \text{Dom}(Q) \cap L^1$. Therefore

$$\|f\|_2^2 \leq 2tQ(f) + c_1^2 t^{-\mu/2}\|f\|_1^2$$

and (ii) follows by putting

$$t = Q(f)^{-2/(\mu+2)}\|f\|_1^{4/(\mu+2)}.$$

Conversely given (ii) we choose $0 \leq f \in L^1 \cap L^2$ and put $u(t) = \|f_t\|_2^2$ where $f_t = e^{-Ht}f$. Then

$$-du/dt = Q(f_t) \geq \|f_t\|_2^{2+4/\mu}/c_2\|f_t\|_1^{4/\mu} \geq u^{1+2/\mu}/c_2\|f\|_1^{4/\mu}.$$

Therefore

$$\frac{d}{dt}(u^{-2/\mu}) \geq \frac{2}{c_2\mu\|f\|_1^{4/\mu}}$$

and

$$u(t)^{-2/\mu} \geq u(t)^{-2/\mu} - u(0)^{-2/\mu}$$
$$\geq 2t/c_2\mu\|f\|_1^{4/\mu}.$$

Therefore

$$\|e^{-Ht}f\|_2 \leq (c_2\mu/2t)^{\mu/4}\|f\|_1$$
$$= c_1 t^{-\mu/4}\|f\|_1.$$

This bound extends to all $0 \leq f \in L^1$ by an approximation argument and then to all $f \in L^1$ by the positivity of e^{-Ht}. Finally (i) follows by duality.

Corollary 2.4.7. *Let e^{-Ht} be a symmetric Markov semigroup on $L^2(\Omega)$ and let $\mu > 0$. Then the following two bounds are equivalent:*

(i)
$$\|e^{-Ht}f\|_\infty \leq c_1 t^{-\mu/4}\|f\|_2$$

for some $c_1 < \infty$, all $0 < t \leq 1$ and all $f \in L^2$.

(ii)
$$\|f\|_2^{2+4/\mu} \leq c_2\{Q(f) + \|f\|_2^2\}\|f\|_1^{4/\mu}$$

for some $c_2 < \infty$ and all $0 \leq f \in \text{Dom}(Q) \cap L^1$.

Proof. If we apply Theorem 2.4.6 to $(H + 1)$ we find that (ii) is equivalent to

(iii)
$$\|e^{-Ht}f\|_\infty \leq c_1 t^{-\mu/4} e^t \|f\|_2$$

for all $0 < t < \infty$ and all $f \in L^2$. It is trivial that (iii) implies (i) and the

converse follows from the inequality
$$\|e^{-Ht}f\|_\infty \leq c_1 \|e^{-H(t-1)}f\|_2 \leq c_1 \|f\|_2$$
valid for all $t \geq 1$.

In conclusion, we mention that it is also possible to use Nash inequalities to investigate situations where $\|e^{-Ht}f\|_\infty/\|f\|_2$ has different power decay as $t \to 0$ and $t \to \infty$ as in Example 2.3.2. However, neither Nash nor Sobolev inequalities can be applied to situations such as that of Example 2.3.4. For concrete examples where this more singular behaviour actually occurs see Section 4.5 and in particular Theorem 4.5.10.

Notes

Section 2.1 The importance of hypercontractivity in constructive quantum field theory was first explained by Nelson (1966); his ideas were followed up rapidly, and are surveyed in Simon (1974). Much of the recent work on hypercontractivity as opposed to ultracontractivity centres upon determining the optimum constant $T(p,q)$ in Theorem 2.1.1. For the harmonic oscillator this was achieved by Nelson (1973), with Gross (1976) giving another proof. For recent work concerning the optimum constant of other operators see Mueller and Weissler (1982), Rothaus (1981; 1985; 1986) and Bakry and Emery (1984; 1985). The relationship between hypercontractivity and logarithmic Sobolev inequalities was first demonstrated by Gross (1976). The importance of his paper was widely appreciated even before publication and led to work by Carmona (1974), Eckmann (1974) and Rosen (1976). The notion of ultracontractivity was not studied until the papers of Davies (1983) and Davies and Simon (1984), the delay being partly caused by the fact that ultracontractive bounds have no role in quantum field theory. Theorems 2.1.4 and 2.1.5 are taken from Davies and Simon (1984) and the remainder of the section is of a standard character.

Section 2.2 The ideas in this section are mainly due to Gross (1976) as adapted by Davies and Simon (1984). For elliptic operators H, Lemma 2.2.6 is trivial but the fact that it holds in complete generality was proved independently by Varopoulos (1985C) and Stroock (1984) p. 183. Theorem 2.2.7 is essentially taken from Davies and Simon (1984).

Section 2.3 Many of the examples given here are to be found in a less complete form in Davies and Simon (1984). For applications of Example 2.3.4 see Theorem 4.5.4, Davies (1987A) and Varopoulos (1985C). The significance of the fact, used in Theorem 2.3.6, that logarithmic Sobolev inequalities are monotone in the quadratic form, has only gradually become obvious.

Section 2.4 Theorem 2.4.2 is due to Varopoulos (1985C), although Fukushima (1977) has a weaker result of a similar character. The ideas behind Theorem 2.4.6 go back to Nash (1958) but in the present form it is due to Fabes and Stroock (1986). See also Carlen, Kusuoka and Stroock (1987) for further details.

3
Gaussian bounds on heat kernels

3.1 Introduction

If Ω is a region in $\mathbb{R}^N$ and H is a strictly elliptic operator statisfying Dirichlet boundary conditions on $L^2(\Omega)$, we have shown that the heat kernel $K(t, x, y)$ of e^{-Ht} satisfies

$$0 \leqslant K(t, x, y) \leqslant c t^{-N/2}. \qquad (3.1.1)$$

In the particular case $H_0 = -\Delta$ we actually have

$$K_0(t, x, y) = (4\pi t)^{-N/2} \exp\{-(x-y)^2/t\} \qquad (3.1.2)$$

so that for this case we see that (3.1.1) is a good estimate when $x = y$ but poor if one is well away from the diagonal. We shall actually establish upper and lower bounds similar to (3.1.2) in the case where H is uniformly elliptic on $L^2(\mathbb{R}^N)$, namely

$$c_1 t^{-N/2} \exp\{-b_1(x-y)^2/t\} \leqslant K(t, x, y) \leqslant c_2 t^{-N/2} \exp\{-b_2(x-y)^2/t\} \qquad (3.1.3)$$

for all $t > 0$ and $x, y \in \mathbb{R}^N$, where b_i and c_i are positive constants. Although this type of bound has been known for some time, much simpler and more powerful methods of proof have recently been discovered and we present these here.

We comment that the upper bound we shall obtain does not depend upon uniform ellipticity of the operator H. Indeed the method gives Gaussian upper bounds for the heat kernels of certain uniformly hypoelliptic operators on $\mathbb{R}^N$.

It is a consequence of Moser's Harnack inequality that the heat kernel of an elliptic operator is strictly positive and continuous in great generality. If the coefficients are smooth then this may be shown along the lines of Theorem 5.2.1. It is not essential for our purposes to refer to Moser's results, since all our bounds on $K(t, x, y)$ can be considered as holding almost everywhere on $\Omega \times \Omega$ for every $t > 0$; by redefining K on a null set in $\Omega \times \Omega$ for every $t > 0$ they then become valid everywhere. We shall not refer to this question explicitly again.

Before proceeding we note that (3.1.3) implies certain bounds for the Green functions, of which the following is typical.

Theorem 3.1.1. *If $N \geq 3$ and the heat kernel of H satisfies (3.1.3), then there exist positive constants a_i such that the Green function G of H^{-1} satisfies*

$$a_1|x - y|^{-(N-2)} \leq G(x, y) \leq a_2|x - y|^{-(N-2)}.$$

3.2. The upper bound

In order to provide a measure of generality we suppose that Ω is a region in $\mathbb{R}^N$ and that H is an elliptic operator satisfying Dirichlet boundary conditions on $L^2(\Omega)$ for which the logarithmic Sobolev inequality

$$\int f^2 \log f \, dx \leq \varepsilon Q(f) + \beta(\varepsilon)\|f\|_2^2 + \|f\|_2^2 \log \|f\|_2 \quad (3.2.1)$$

holds for all $0 \leq f \in \text{Quad}(H) \cap L^1 \cap L^\infty$, all $\varepsilon > 0$, and some function $\beta(\varepsilon)$. If H is strictly elliptic then (3.1.1) holds by Theorem 2.3.6, and (3.2.1) holds with

$$\beta(\varepsilon) = c - (N/4)\log \varepsilon$$

but we do not wish to restrict attention to this particular case.

The core of our estimates is the following lemma, which will allow us to prove weighted ultracontractive bounds. We put $\phi = e^{\alpha \psi}$ where $\alpha \in \mathbb{R}$ and $\psi: \Omega \to \mathbb{R}$ is a C^∞ bounded function which satisfies

$$\sum a_{ij}(x)\frac{\partial \psi}{\partial x_i}\frac{\partial \psi}{\partial x_j} \leq 1$$

everywhere on Ω, where

$$Q(f) = \int_\Omega \sum a_{ij}(x)\frac{\partial f}{\partial x_i}\frac{\partial \bar{f}}{\partial x_j}\, dx$$

for all $f \in C_c^\infty(\Omega)$.

Lemma 3.2.1. *If $0 \leq f \in C_c^\infty(\Omega)$ and $2 \leq p < \infty$ and $0 < \mu < 1$ then*

$$(1 - \mu)\langle H^{\frac{1}{2}}f, H^{\frac{1}{2}}f^{p-1} \rangle \leq \langle H^{\frac{1}{2}}(\phi f), H^{\frac{1}{2}}(\phi^{-1}f^{p-1}) \rangle$$
$$+ \alpha^2(1 + (p-2)^2/4(p-1)\mu)\|f\|_p^p.$$

Proof. We have

$$\langle H^{\frac{1}{2}}(\phi f), H^{\frac{1}{2}}(\phi^{-1}f^{p-1}) \rangle = \int_\Omega \sum a_{ij}\left(f\frac{\partial \phi}{\partial x_i} + \phi \frac{\partial f}{\partial x_i}\right)$$
$$\times \left\{-\phi^{-2}f^{p-1}\frac{\partial \phi}{\partial x_j} + (p-1)\phi^{-1}f^{p-2}\frac{\partial f}{\partial x_j}\right\}dx$$

$$= \int_\Omega \sum a_{ij} \left\{ -\alpha^2 f^p \frac{\partial \psi}{\partial x_i} \frac{\partial \psi}{\partial x_j} \right.$$
$$+ \alpha(p-1) f^{p-1} \frac{\partial f}{\partial x_i} \frac{\partial \psi}{\partial x_j} - \alpha f^{p-1} \frac{\partial f}{\partial x_i} \frac{\partial \psi}{\partial x_j}$$
$$\left. + (p-1) f^{p-2} \frac{\partial f}{\partial x_i} \frac{\partial f}{\partial x_j} \right\} dx$$
$$\geq -\alpha^2 \int_\Omega f^p \, dx$$
$$- |\alpha|(p-2) \int_\Omega f^{p-1} \left(\sum a_{ij} \frac{\partial f}{\partial x_i} \frac{\partial f}{\partial x_j} \right)^{\frac{1}{2}} dx$$
$$+ \int_\Omega \sum a_{ij} \frac{\partial f}{\partial x_i} \frac{\partial f^{p-1}}{\partial x_j} \, dx.$$

Also if $0 < s < \infty$ then

$$\int_\Omega f^{p-1} \left(\sum a_{ij} \frac{\partial f}{\partial x_i} \frac{\partial f}{\partial x_j} \right)^{\frac{1}{2}} dx = \frac{2}{p} \int_\Omega f^{p/2} \left(\sum a_{ij} \frac{\partial f^{p/2}}{\partial x_i} \frac{\partial f^{p/2}}{\partial x_j} \right)^{\frac{1}{2}} dx$$
$$\leq \frac{s}{p} \int_\Omega \sum a_{ij} \frac{\partial f^{p/2}}{\partial x_i} \frac{\partial f^{p/2}}{\partial x_j} \, dx + \frac{1}{sp} \int f^p \, dx$$
$$= \frac{sp}{4(p-1)} \int_\Omega \sum a_{ij} \frac{\partial f}{\partial x_i} \frac{\partial f^{p-1}}{\partial x_j} \, dx + \frac{1}{sp} \int f^p \, dx$$
$$= \frac{sp}{4(p-1)} \langle H^{\frac{1}{2}} f, H^{\frac{1}{2}} f^{p-1} \rangle + \frac{1}{sp} \|f\|_p^p.$$

Therefore

$$\langle H^{\frac{1}{2}}(\phi f), H^{\frac{1}{2}}(\phi^{-1} f^{p-1}) \rangle \geq \langle H^{\frac{1}{2}} f, H^{\frac{1}{2}} f^{p-1} \rangle - \alpha^2 \|f\|_p^p$$
$$- |\alpha|(p-2) \left\{ \frac{sp}{4(p-1)} \langle H^{\frac{1}{2}} f, H^{\frac{1}{2}} f^{p-1} \rangle + \frac{1}{sp} \|f\|_p^p \right\}.$$

If we put

$$s = \frac{4(p-1)\mu}{|\alpha| p(p-2)}$$

then we obtain the bound of the lemma.

Corollary 3.2.2. *If* $0 \leq f \in C_c^\infty(\Omega), 0 < \mu < 1, 0 < \varepsilon < \infty$ *and* $2 < p < \infty$ *then*

$$\int_\Omega f^p \log f \, dx \leq \varepsilon \langle H^{\frac{1}{2}}(\phi f), H^{\frac{1}{2}}(\phi^{-1} f^{p-1}) \rangle + \gamma(\varepsilon, p) \|f\|_p^p$$
$$+ \|f\|_p^p \log \|f\|_p. \tag{3.2.2}$$

where

$$\gamma(\varepsilon, p) = 2\beta\{\varepsilon(1-\mu)\}p^{-1} + \varepsilon\alpha^2\{1 + (p-2)^2/4(p-1)\mu\}.$$

Proof. We replace f by $f^{p/2}$ in (3.2.1) as in Lemma 2.2.6 to obtain

$$\int f^p \log f \, dx \leq \varepsilon(1-\mu)\langle H^{\frac{1}{2}}f, H^{\frac{1}{2}}f^{p-1}\rangle + 2\beta(\varepsilon(1-\mu))p^{-1}\|f\|_p^p$$
$$+ \|f\|_p^p \log \|f\|_p$$

and then apply Lemma 3.2.1.

Our next goal is to integrate (3.2.2) to obtain ultracontractive bounds on $\exp(-\phi^{-1}H\phi t)$, but we first need to extend (3.2.2) to a larger class of f. We put

$$\mathscr{D}_1 = \bigcup_{t>0} e^{-Ht}(L^1 \cap L^\infty)_+$$

and

$$\mathscr{D} = \phi^{-1}\mathscr{D}_1.$$

Since $\phi^{\pm 1}$ are bounded functions, they define bounded and invertible multiplication operators on L^p for $1 \leq p \leq \infty$. We define $K = \phi^{-1}H\phi$ so that

$$\text{Dom}(K) = \{f : \phi f \in \text{Dom}(H)\}.$$

Then $\mathscr{D}$ is dense in L^p_+ for all $1 \leq p < \infty$, and $\mathscr{D}$ is invariant under the semigroup e^{-Kt}. Since e^{-Ht} is holomorphic on L^p for all $1 < p < \infty$ by Theorem 1.4.2, the same holds for e^{-Kt}. In particular $\mathscr{D}$ lies in the domain of K_p^n for all n, where K_p is the generator of e^{-Kt} on L^p for $1 < p < \infty$. We now have to prove that (3.2.2) holds for all $0 \leq f \in \mathscr{D}$, or equivalently that

$$\int_\Omega (\phi^{-1}g)^p \log(\phi^{-1}g) \, dx \leq \varepsilon \langle H^{\frac{1}{2}}g, H^{\frac{1}{2}}(\phi^{-p}g^{p-1})\rangle + \gamma(\varepsilon, p)\|\phi^{-1}g\|_p^p$$
$$+ \|\phi^{-1}g\|_p^p \log \|\phi^{-1}g\|_p \quad (3.2.3)$$

for all $0 \leq g \in \mathscr{D}_1$, starting from the fact that this holds for all $0 \leq g \in C_c^\infty$.

We note that $\mathscr{D}_1 \subseteq \mathscr{D}_2$ where

$$\mathscr{D}_2 = \{\text{Quad}(H) \cap L^1 \cap L^\infty\}_+.$$

We shall use the inner product

$$\langle f, g \rangle' = \langle f, g \rangle + \langle H^{\frac{1}{2}}f, H^{\frac{1}{2}}g \rangle$$

on Quad(H) and the associated complete norm

$$\|\|f\|\| = \{\langle f, f \rangle'\}^{\frac{1}{2}}.$$

We say that $f_n \xrightarrow{w} f$ in Quad(H) if

$$\langle f_n, g \rangle' \to \langle f, g \rangle'$$

for all $g \in \text{Quad}(H)$. A sufficient condition for this is that $|||f_n|||$ is uniformly bounded and $||f_n - f|| \to 0$.

Lemma 3.2.3. *The bound (3.2.2) is valid for all $0 \leq f \in \mathcal{D}_1$, $0 < \varepsilon < \infty$ and $2 < p < \infty$.*

Proof. By the above comments it is sufficient to show that (3.2.3) is valid for all $0 \leq g \in \mathcal{D}_2$. We first confirm that (3.2.3) makes sense. Since $g \in L^1 \cap L^\infty$, $\phi^{-1}g \in L^1 \cap L^\infty$ and the LHS of (3.2.3) is finite. If $g \in \mathcal{D}_2$ and $p > 2$ then

$$g^{p-1} = F(g)$$

where F has bounded derivative and vanishes at 0. It follows from Theorem 1.3.3 that

$$Q(g^{p-1}) \leq cQ(g) < \infty. \qquad (3.2.4)$$

We deduce that $g \in \mathcal{D}_2$ implies $g^{p-1} \in \mathcal{D}_2$. The finiteness of the RHS of (3.2.3) now follows if we show that $\mathcal{D}_2$ is invariant under multiplication by ϕ^{-p}. Since ϕ^{-1} is bounded it is sufficient to show that $f \in \text{Quad}(H)$ implies $\phi^{-p}f \in \text{Quad}(H)$. This results from the fact that $|||f|||$ and $|||\phi^{-p}f|||$ are equivalent norms on the dense subspace $C_c^\infty(\Omega)$ of $\text{Quad}(H)$ by an easy computation.

We are now ready to produce our approximation procedure. If $0 \leq g \in \mathcal{D}_2$ then there exists a sequence $h_n \in C_c^\infty$ such that $|||h_n - g||| \to 0$. If $||g||_\infty = k$ then there exist $F_m \in C^\infty$ with $0 \leq F_n \leq 2k$, $0 \leq F'_n \leq 1$ and $F_n(0) = 0$ such that $g_n = F_n(h_n)$ satisfy $g_n \in C_c^\infty$, $0 \leq g_n \leq 2k$, $||g_n - g||^2 \to 0$ and $Q(g_n) \leq Q(g)$. Since these bounds imply

$$\limsup |||g_n||| \leq |||g|||$$

we deduce that

$$|||g_n - g||| \to 0. \qquad (3.2.5)$$

Secondly we see as in the proof of (3.2.4) that

$$Q(g_n^{p-1}) \leq cQ(g_n) \leq cQ(g)$$

for all n. Using the fact that the multiplication by ϕ^{-p} is a bounded operator on $\text{Quad}(H)$ we deduce that

$$|||\phi^{-p}g_n^{p-1}||| \leq d < \infty$$

for all n. This implies that

$$\phi^{-p}g_n^{p-1} \xrightarrow{w} \phi^{-p}g^{p-1}. \qquad (3.2.6)$$

By combining (3.2.5) and (3.2.6) we discover that

$$\langle H^{\frac{1}{2}}g_n, H^{\frac{1}{2}}(\phi^{-p}g_n^{p-1})\rangle \to \langle H^{\frac{1}{2}}g \cdot H^{\frac{1}{2}}(\phi^{-p}g^{p-1})\rangle.$$

We finally have to control the convergence of the other terms in (3.2.3). Since $\|g_n - g\|_2 \to 0$ and $0 \leq g_n \leq 2k$ for all n it follows that $\|g_n - g\|_p \to 0$. Fatou's lemma implies that

$$\int (\phi^{-1}g)^p \log_+ (\phi^{-1}g) \, dx \leq \liminf_{n \to \infty} \int (\phi^{-1}g_n)^p \log_+ (\phi^{-1}g_n) \, dx$$

and the proof that

$$\int (\phi^{-1}g)^p \log_- (\phi^{-1}g) \, dx = \lim_{n \to \infty} \int (\phi^{-1}g_n)^p \log_- (\phi^{-1}g) \, dx$$

depends upon the fact that $s^{p-2} \log_- (s)$ is a bounded function for any $p > 2$, while $\phi^{-1}g_n$ converges to $\phi^{-1}g$ in L^2 norm.

Theorem 3.2.4. *Let $\varepsilon(p) > 0$ be a continuous function defined for $2 < p < \infty$ and put*

$$\Gamma(p) = \gamma(\varepsilon(p), p)$$
$$= 2\beta\{\varepsilon(p)(1-\mu)\}p^{-1} + \varepsilon(p)\alpha^2\{1 + (p-2)^2/4(p-1)\mu\}.$$

If

$$t = \int_2^\infty p^{-1}\varepsilon(p) \, dp, \qquad M = \int_2^\infty p^{-1}\Gamma(p) \, dp$$

are both finite then $\exp(-\phi^{-1}H\phi t)$ maps L^2 to L^∞ and

$$\|\exp(-\phi^{-1}H\phi t)\|_{\infty,2} \leq e^M.$$

Proof. If $K = \phi^{-1}H\phi$ then we have shown that

$$\int_\Omega f^p \log f \, dx \leq \varepsilon \langle Kf, f^{p-1} \rangle + \gamma(\varepsilon, p)\|f\|_p^p + \|f\|_p^p \log \|f\|_p$$

for all $0 \leq f \in \mathcal{D} \subseteq \text{Dom}(K)$. For the remainder of the proof we may follow Theorem 2.2.7; the failure of K to be self-adjoint does not matter in the proof.

The application of the above theorem depends upon making suitable choices for $\varepsilon(p)$ and μ, depending upon the given function $\beta(\varepsilon)$. The following is the most straightforward case.

Theorem 3.2.5. *If*

$$\beta(\varepsilon) = c - (N/4)\log \varepsilon \qquad (3.2.7)$$

then

$$\|\phi^{-1}e^{-Ht}\phi\|_{\infty,2} \leq a_\delta t^{-N/4} e^{(1+\delta)\alpha^2 t} \qquad (3.2.8)$$

for all $0 < t < \infty$, and all $0 < \delta < 1$.

Proof. We put $\mu = \frac{1}{2}$ and
$$\varepsilon(p) = \lambda 2^\lambda t p^{-\lambda}$$
where $\lambda > 1$, so that
$$t = \int_2^\infty p^{-1} \varepsilon(p) \, dp.$$
Theorem 3.2.4 then yields
$$M(t) = \int_2^\infty \frac{2}{p^2} \left\{ c - \frac{N}{4} \log(\lambda 2^{\lambda-1} t p^{-\lambda}) \right\} dp$$
$$+ \int_2^\infty \lambda 2^\lambda t p^{-\lambda} \alpha^2 \{ 1 + (p-2)^2/2(p-1) \} p^{-1} \, dp.$$
Now $p \geq 2$ implies
$$1 + (p-2)^2/2(p-1) \leq p/2$$
so
$$M(t) \leq -\frac{N}{4} \log t + \frac{\lambda}{\lambda - 1} \alpha^2 t + c_\lambda.$$
We finally choose λ so that
$$\frac{\lambda}{\lambda - 1} = 1 + \delta.$$

Corollary 3.2.6. *Given (3.2.7) we have*
$$0 \leq K(t, x, y) \leq c_\delta t^{-N/2} \exp(-\{\psi(x) - \psi(y)\}^2 / 4(1 + \delta)t)$$
for all $0 < t < \infty$, $0 < \delta < 1$ and $x, y \in \Omega$.

Proof. Replacing α by $-\alpha$ and taking adjoints of (3.2.8) we obtain
$$\|\phi^{-1} e^{-Ht} \phi\|_{2,1} \leq a_\delta t^{-N/4} e^{(1+\delta)\alpha^2 t}$$
and multiplying the two bounds yields
$$\|\phi^{-1} e^{-Ht} \phi\|_{\infty,1} \leq c_\delta t^{-N/2} e^{(1+\delta)\alpha^2 t}.$$
This is equivalent to the bound
$$0 \leq \phi(x)^{-1} K(t, x, y) \phi(y) \leq c_\delta t^{-N/2} e^{(1+\delta)\alpha^2 t}$$
and hence to
$$0 \leq K(t, x, y) \leq c_\delta t^{-N/2} \exp\{(1+\delta)\alpha^2 t + \alpha(\psi(x) - \psi(y))\}.$$
We now make the choice
$$\alpha = \frac{\psi(y) - \psi(x)}{2(1+\delta)t}$$
to complete the proof.

Theorem 3.2.7. *If (3.2.7) holds and we define the metric d on Ω by*

$$d(x,y) = \sup\left\{|\psi(x) - \psi(y)| : \psi \text{ is } C^\infty \text{ and bounded with} \right.$$
$$\left. \sum a_{ij}\frac{\partial \psi}{\partial x_i}\frac{\partial \psi}{\partial x_j} \leq 1 \text{ on } \Omega\right\} \quad (3.2.9)$$

then the heat kernel K of e^{-Ht} *satisfies*

$$0 \leq K(t,x,y) \leq c_\delta t^{-N/2} \exp\{-d(x,y)^2/4(1+\delta)t\}$$

for all $0 < t < \infty$, $0 < \delta < 1$, *and* $x, y \in \Omega$.

Proof. This follows directly from Corollary 3.2.6.

The above theorem is applicable to all strictly elliptic operators H on $L^2(\Omega)$ with Dirichlet boundary conditions. The metric d on Ω is determined geometrically by (3.2.9) and need not be equivalent to the Euclidean metric. The following special case is, however, worth recording.

Corollary 3.2.8. *If Ω is a region in $\mathbb{R}^N$ and*

$$Hf = -\sum \frac{\partial}{\partial x_i}\left\{a_{ij}(x)\frac{\partial f}{\partial x_j}\right\}$$

with Dirichlet boundary conditions where

$$0 < \lambda \leq a(x) \leq \mu < \infty$$

for all $x \in \Omega$ *then*

$$0 \leq K(t,x,y) \leq c_{\delta,\lambda} t^{-N/2} \exp\{-(x-y)^2/4(1+\delta)\mu t\}$$

for all $0 < t < \infty$, $0 < \delta < 1$ *and* $x, y \in \Omega$.

Proof. We see from (3.2.9) that

$$d(x,y) \geq \sup\left\{|\psi(x) - \psi(y)| : \psi \text{ is } C^\infty \text{ and bounded with} \right.$$
$$\left. \sum \mu \delta_{ij}\frac{\partial \psi}{\partial x_i}\frac{\partial \psi}{\partial x_j} \leq 1 \text{ on } \Omega\right\}$$
$$= \sup\{|\psi(x) - \psi(y)| : |\nabla \psi| \leq \mu^{-\frac{1}{2}}\}$$
$$= |x-y|\mu^{-\frac{1}{2}}.$$

All of the above results can be modified to deal with the case of Neumann boundary conditions. However, if Ω is bounded and H is an elliptic differential operator with Neumann boundary conditions, then $H1 = 0$, so $0 \in \text{Sp}(H)$ and the asymptotic behaviour as $t \to \infty$ will not be the same as in the case of Dirichlet boundary conditions.

Theorem 3.2.9. *Let Ω be a bounded region in $\mathbb{R}^N$ with the extension property and let*

$$Hf = -\sum \frac{\partial}{\partial x_i}\left\{a_{ij}(x)\frac{\partial f}{\partial x_j}\right\}$$

act on $L^2(\Omega)$ with Neumann boundary conditions, where

$$0 < \lambda \leq a(x) \leq \mu < \infty$$

for all $x \in \Omega$. Then

$$0 \leq K(t,x,y) \leq c_{\delta,\lambda,\Omega}(t^{-N/2} \vee 1)\exp\{-(x-y)^2/4(1+\delta)\mu t\}$$

for all $0 < t < \infty$, $0 < \delta < 1$ and $x,y \in \Omega$.

Proof. If $t \geq 1$ then the RHS is bounded below since Ω is a bounded region. Since $\|e^{-Ht}\|_{\infty,1}$ is monotonically decreasing it is therefore sufficient to treat the case $0 < t \leq 1$.

For such t we proved in Theorem 2.4.4 that

$$0 \leq K(t,x,y) \leq ct^{-N/2}$$

and in Lemma 1.7.11 that $C_c^\infty(\bar{\Omega})$ is a form core for $W^{1,2}(\Omega) = \text{Dom}(H^{\frac{1}{2}})$. We may now repeat the calculations in this section with only minor modifications.

Note 3.2.10. In both Theorem 3.2.7 and Theorem 3.2.9 a spurious constant $\delta > 0$ appears. If one examines the arguments more carefully one can find the precise dependence of c_δ upon δ, and obtain a sharper upper bound on the heat kernel by optimising over δ. We leave the reader to fill in the details.

Example 3.2.11. Let $a(x) \geq 0$ be a smooth matrix-valued function on $\mathbb{R}^N$ and define the subelliptic operator $H: C_c^\infty \to C_c^\infty$ by

$$Hf = -\sum \frac{\partial}{\partial x_i}\left(a_{ij}\frac{\partial f}{\partial x_j}\right).$$

We extended H to a self-adjoint operator by Theorem 1.2.5. Then e^{-Ht} is a symmetric Markov semigroup on $L^2(\mathbb{R}^N)$ by Theorem 1.3.5, whose proof does not use the ellipticity assumption. We now make the extra assumption that the heat kernel exists and satisfies

$$0 \leq K(t,x,y) \leq at^{-\mu/2}$$

for some $a > 0$, $\mu \geq N$ and all $0 < t \leq 1$ and $x,y \in \mathbb{R}^N$. This type of bound is highly non-trivial but follows if H satisfies Hörmander's hypoellipticity condition uniformly with respect to x in a suitable sense.

Under this condition Theorem 2.2.3 implies that (3.2.1) holds for all $0 < \varepsilon < \infty$ with

$$\beta(\varepsilon) = c - (\mu/4)\log\varepsilon + \varepsilon.$$

If the metric d associated to the hypoelliptic operator H is defined by (3.2.9) then the calculations of this section, with the modified choice of $\beta(\varepsilon)$, lead to

$$0 \leq K(t,x,y) \leq c_\delta t^{-\mu/2} \exp\{-d(x,y)^2/4(1+\delta)t\}$$

for all $0 < t \leq 1$, $0 < \delta < 1$ and $x, y \in \mathbb{R}^N$.

For large t the behaviour of the heat kernel is rather different, and indeed one finds that

$$K(t,x,y) \sim ct^{-N/2}$$

as $t \to \infty$. We shall not, however, pursue this.

3.3 The lower bound

Throughout this section we assume that the elliptic operator H is defined in the region $\Omega \subseteq \mathbb{R}^N$ by

$$Hf = -\sum \frac{\partial}{\partial x_i}\left\{a_{ij}(x)\frac{\partial f}{\partial x_j}\right\} \qquad (3.3.1)$$

subject to Dirichlet boundary conditions, and that

$$0 < \lambda \leq a(x) \leq \mu < \infty$$

for all $x \in \Omega$. By Corollary 3.2.8 we therefore have a bound on the heat kernel of the form

$$0 \leq K(t,x,y) \leq ct^{-N/2} \exp\{-b(x-y)^2/t\}$$

for all $t > 0$ and $x, y \in \Omega$, where b and c are positive constants. Our task in this section is to obtain a similar lower bound on the heat kernel.

Let $B \subseteq \Omega$ be any ball, with centre b and radius $\beta > 0$, and let ϕ be the ground state of $-\Delta$ in $L^2(B)$ subject to Dirichlet boundary conditions. We normalise ϕ by $\|\phi\|_2 = 1$ and observe that

$$0 \leq \phi \in W_0^{1,2}(B) \subseteq W_0^{1,2}(\Omega).$$

Now ϕ may be computed explicitly by separation of variables in polar coordinates, and this leads to the bounds

$$|x-b| \leq \beta \Rightarrow 0 \leq \phi(x) \leq c_1 \beta^{-N/2}, \quad |x-b| \leq \beta/2 \Rightarrow \phi(x) \geq c_2 \beta^{-N/2}$$

where c_1 and c_2 are positive constants. The smallest two eigenvalues of $-\Delta$ on $L^2(B)$ are $E\beta^{-2}$ and $F\beta^{-2}$ where $0 < E < F$ and E, F are again explicitly computable by separation of variables. Finally let ε be a constant and f a function on B such that

$$0 < \varepsilon < \beta^{-N}, f \geq 0, \int f \, dx = 1$$

and

$$\operatorname{supp}(f) \subseteq \{x : |x-b| < \beta/4\}.$$

It will be of the greatest importance that c, c_1, c_2, E, F and other constants below are independent of B, b, β, f, Ω and ε provided these satisfy the above conditions. We put

$$u_t(x) = (e^{-Ht}f)(x) = \int_\Omega K(t,x,y)f(y)\,dy$$

and will be concerned to obtain lower bounds on u_t. The idea behind the proof of our first lemma is that e^{-Ht} is approximately probability-conserving for short times.

Lemma 3.3.1. *There exist constants $c_3 > 0$ and $T_3 > 0$ such that $0 < t < T_3 \beta^2$ implies*

$$\int \phi^2 u_t \, dx \geq c_3 \beta^{-N}.$$

Proof. Let B_r be the ball with centre b and radius $r\beta/4$, let χ be the characteristic function of $B_4 \setminus B_3$, and let H' be the operator in $L^2(B_4)$ associated with (3.3.1) and Neumann boundary conditions. If $\alpha > 0$ and $t > 0$ then

$$e^{-H't}f = \exp\{-(H'+\alpha\chi)t\}f + \int_{s=0}^{t} e^{-H's}\alpha\chi \exp\{-(H'+\alpha\chi)(t-s)\}f\,ds$$

$$= \exp\{-(H'+\alpha\chi)t\}f + \int_{s=0}^{t} e^{-H's}g_s\,ds \quad (3.3.2)$$

where $g_s \geq 0$ and $\mathrm{supp}(g_s) \subseteq B_4 \setminus B_3$. Since $e^{-H't}$ conserves probability we see that

$$0 \leq \int_{x\in B_4}\int_{s=0}^{t} g_s(x)\,ds\,dx \leq 1.$$

We next observe that the heat kernel $K'(t,x,y)$ of $e^{-H't}$ satisfies

$$0 \leq K'(t,x,y) \leq c't^{-N/2}\exp\{-b'(x-y)^2/t\}$$

where $c' > 0$ and $b' > 0$, by Theorem 3.2.9 provided t lies in the stated range. By using this bound we discover that

$$\int_{B_2} (e^{-H't}f)(x)\,dx = 1 - \int_{B_4\setminus B_2} (e^{-H't}f)(x)\,dx$$

$$= 1 - \int_{x\in B_4\setminus B_2}\int_{y\in B_1} K'(t,x,y)f(y)\,dy\,dx$$

$$\geq \tfrac{2}{3}$$

provided $0 < t \leq T_{3,1}\beta^2$. Also

$$\int_{s=0}^{t}\int_{x\in B_2}(e^{-H's}g_s)(x)\,dx\,ds = \int_{s=0}^{t}\int_{x\in B_2}\int_{y\in B_4\setminus B_3}K'(s,x,y)g_s(y)\,dy\,dx\,ds$$
$$\leq \tfrac{1}{3}$$

provided $0 < t \leq T_{3,2}\beta^2$. Integrating both sides of (3.3.2) over B_2 now yields

$$\int_{B_2}(\exp\{-(H'+\alpha\chi)t\}f)(x)\,dx \geq \tfrac{1}{3} \qquad (3.3.3)$$

provided $0 < t \leq T_3\beta^2$. If H'' denotes the operator on $L^2(B_3)$ associated with (3.3.1) and Dirichlet boundary conditions, then letting $\alpha \to +\infty$ in (3.3.3) yields

$$\int_{B_2}(e^{-H''t}f)(x)\,dx \geq \tfrac{1}{3}.$$

But $f \geq 0$ so

$$e^{-Ht}f \geq e^{-H''t}f$$

by Theorem 2.1.6. Therefore

$$\int \phi^2 u_t\,dx \geq c_2^2\beta^{-N}\int_{B_2}u_t\,dx$$
$$\geq c_2^2\beta^{-N}\int_{B_2}(e^{-H''t}f)(x)\,dx$$
$$\geq \tfrac{1}{3}c_2^2\beta^{-N}.$$

We next obtain a lower bound on the function

$$G(t) = \int \phi^2 \log\{t^{N/2}(u_t+\varepsilon)/2c\}\,dx \qquad (3.3.4)$$

for small enough $t > 0$. We note that

$$t^{N/2}\varepsilon/2c \leq t^{N/2}(u_t+\varepsilon)/2c$$
$$\leq t^{N/2}(ct^{-N/2}+\beta^{-N})/2c$$
$$\leq \tfrac{1}{2}+t^{N/2}/2c\beta^N < 1$$

provided $0 < t \leq T_1\beta^2$. Therefore

$$\log(\varepsilon/2c) - (N/2)\log t \leq G(t) < 0$$

for all such t. The intuition behind the following calculations is that if

$$u_t \sim c_0 t^{-N/2}\exp\{-b_0(x-b)^2/t\}$$

then

$$G(t) \sim -t^{-1}\beta^2$$

so
$$G'(t) \sim G(t)^2 \beta^{-2}.$$
We actually prove a differential inequality of this type for small $t > 0$.

Lemma 3.3.2. *There exist constants $T_4 > 0$ and $c_4 > 0$ such that*
$$G(T^4 \beta^2) \geq -c_4.$$

Proof. It follows from the definition (3.3.4) of G that
$$G'(t) = \frac{N}{2t} - \int \phi^2 \frac{Hu}{u+\varepsilon} dx$$
$$= \frac{N}{2t} - \int \sum a_{ij} \frac{\partial u}{\partial x_i} \frac{\partial}{\partial x_j}\left(\frac{\phi^2}{u+\varepsilon}\right) dx$$
$$= \frac{N}{2t} + I$$

where
$$I = \int \sum a_{ij} \frac{\partial u}{\partial x_i} \left\{ \frac{\phi^2}{(u+\varepsilon)^2} \frac{\partial u}{\partial x_j} - \frac{2\phi}{u+\varepsilon} \frac{\partial \phi}{\partial x_j} \right\} dx$$
$$= \int \sum a_{ij} \left\{ \frac{\phi}{2} \frac{\partial \log(u+\varepsilon)}{\partial x_i} - 2\frac{\partial \phi}{\partial x_i} \right\} \left\{ \frac{\phi}{2} \frac{\partial \log(u+\varepsilon)}{\partial x_j} - 2\frac{\partial \phi}{\partial x_j} \right\} dx$$
$$+ \frac{3}{4} \int \sum a_{ij} \phi^2 \frac{\partial \log(u+\varepsilon)}{\partial x_i} \frac{\partial \log(u+\varepsilon)}{\partial x_j} dx - 4 \int \sum a_{ij} \frac{\partial \phi}{\partial x_i} \frac{\partial \phi}{\partial x_j} dx$$
$$\geq \frac{3\lambda}{4} \int \phi^2 |\nabla \log(u+\varepsilon)|^2 dx - 4\mu \int |\nabla \phi|^2 dx.$$

We deduce that
$$G'(t) \geq \frac{3\lambda}{4} \int \phi^2 |\nabla f|^2 dx$$

provided
$$h = \log\{t^{N/2} \log(u+\varepsilon)/2c\}$$
and
$$N/4t \geq 4\mu \int (\nabla \phi)^2 dx = 4\mu E \beta^{-2}. \qquad (3.3.5)$$

The inequality (3.3.5) is of the form
$$0 < t \leq T_{4,1} \beta^2.$$
Now
$$\phi h \in W_0^{1,2}(B) = \text{Quad}(-\Delta|_B)$$

and

$$\int |\nabla(\phi h)|^2 \, dx - E\beta^{-2} \int \phi^2 h^2 \, dx$$

$$= \int (\phi \nabla h + h \nabla \phi) \cdot (\phi \nabla h + h \nabla \phi) \, dx - E\beta^{-2} \int \phi^2 h^2 \, dx$$

$$= \int \nabla \phi \cdot \nabla(\phi h^2) \, dx + \int \phi^2 (\nabla h)^2 \, dx - E\beta^{-2} \int \phi^2 h^2 \, dx$$

$$= \int \phi^2 (\nabla h)^2 \, dx$$

so we see by an application of the spectral theorem for $-\Delta|_B$ that

$$\int \phi^2 (\nabla h)^2 \, dx \geq (F-E)\beta^{-2} \| \phi h - \langle \phi h, \phi \rangle \phi \|^2$$

$$= (F-E)\beta^{-2} \int \phi^2 (h - \langle \phi h, \phi \rangle)^2 \, dx$$

$$= (F-E)\beta^{-2} \int \phi^2 \{h - G(t)\}^2 \, dx.$$

Therefore $0 < t \leq T_{4,1}\beta^2$ implies

$$G'(t) \geq \frac{3\lambda(F-E)}{4\beta^2} \int \phi^2 \{h - G(t)\}^2 \, dx \geq 0. \qquad (3.3.6)$$

We next recall that $h \leq 0$. If

$$2 + G(t) \leq h(x,t) \leq 0$$

at some $x \in B$ then

$$1 - \frac{h}{G(t)} \geq 1 - \frac{h}{h-2} = \frac{2}{2-h} \geq e^{h/2}$$

so

$$\{h - G(t)\}^2 \geq G(t)^2 e^h.$$

We deduce from (3.3.6) that

$$G'(t) \geq \frac{3\lambda(F-E)}{4\beta^2} G(t)^2 \int_{h \geq 2 + G(t)} \phi^2 e^h \, dx$$

$$\geq \frac{3\lambda(F-E)}{4\beta^2} G(t)^2 \left\{ \int \phi^2 e^h \, dx - e^{2 + G(t)} \right\}$$

$$\geq \frac{3\lambda(F-E)}{4\beta^2} G(t)^2 \left\{ \frac{t^{N/2}}{2c} \int \phi^2 u \, dx - e^{2 + G(t)} \right\}.$$

If $T_4 = T_3 \wedge T_{4,1}$ then Lemma 3.3.1 implies

$$G'(t) \geq \frac{3\lambda(F-E)}{4\beta^2} G(t)^2 \left\{ \frac{c_3}{2c} t^{N/2} \beta^{-N} - e^{2 + G(t)} \right\}$$

wherever $0 < t \leq T_4\beta^2$. Since $G(t)$ is monotonically increasing by (3.3.6) we deduce that either

$$e^{2+G(T_4\beta^2)} \geq (c_3/4c)(T_4\beta^2/2)^{N/2}\beta^{-N} \qquad (3.3.7)$$

or

$$G'(t) \geq \frac{3\lambda(F-E)}{4\beta^2} G(t)^2 \frac{c_3}{4c}\left(\frac{T_4\beta^2}{2}\right)^{N/2} \beta^{-N} \qquad (3.3.8)$$

for all $T_4\beta^2/2 \leq t \leq T_4\beta^2$. The case (3.3.7) yields

$$G(T_4\beta^2) \geq -c_{4,1} \qquad (3.3.9)$$

while the case (3.3.8) yields

$$G'(t) \geq \beta^{-2} k G(t)^2$$

for some $k > 0$ and all t in the stated interval. Hence

$$\frac{d}{dt}\left\{-G^{-1}(t)\right\} \geq \beta^{-2} k$$

and

$$-G^{-1}(T_4\beta^2) \geq \beta^{-2} k T_4\beta^2/2$$

so

$$G(T_4\beta^2) \geq -2/kT_4 = -c_{4,2}. \qquad (3.3.10)$$

The lemma follows by combining (3.3.9) and (3.3.10).

Lemma 3.3.3. *There exist constants $T_5 > 0$ and $c_5 > 0$ such that $|x-b| < \beta/4$ and $|y-b| < \beta/4$ imply*

$$K(T_5\beta^2, x, y) \geq c_5\beta^{-N}.$$

Proof. Let $u_{1,t}$ and $u_{2,t}$ be two functions constructed as above from f_1 and f_2, both with support in $\{x : |x-b| < \beta/4\}$. If $t = T_4\beta^2$ then

$$I \equiv \int K(2T_4\beta^2, x, y) f_1(x) f_2(y) \, dx \, dy$$

$$= \langle e^{-Ht} f_1, e^{-Ht} f_2 \rangle$$

$$\geq c_1^{-2} \beta^N \int \phi^2 u_{1,t} u_{2,t} \, dx$$

$$= c_{5,1} \beta^{-N} \int \phi^2 (t^{N/2} u_1/2c)(t^{N/2} u_2/2c) \, dx.$$

Therefore

$$\log(c_{5,1}^{-1} \beta^N I) \geq \log \int \phi^2 (t^{N/2} u_1/2c)(t^{N/2} u_2/2c) \, dx$$

$$\geq \int \phi^2 \log\{t^{N/2} u_1/2c\} \, dx + \int \phi^2 \log\{t^{N/2} u_2/2c\} \, dx$$

$$\geq -2c_4$$

The lower bound

by applying Lemma 3.3.2 after taking the limit $\varepsilon \to 0$. We conclude that

$$I \geq c_{5,1} \beta^{-N} e^{-2c_4}$$

which yields the required result if we put $T_5 = 2T_4$ and use the fact that f_i are arbitrary subject to $f_i = 0$, $\int f_i = 1$ and

$$\text{supp}(f_i) \subseteq \{x : |x - b| < \beta/4\}.$$

We are now able to obtain our global lower bounds on the heat kernel. The easiest case to deal with is when $\Omega = \mathbb{R}^N$.

Theorem 3.3.4. *Let H be a uniformly elliptic operator acting on $L^2(\mathbb{R}^N)$. Then there exist constants $c > 0$ and $a > 0$ such that the heat kernel $K(t, x, y)$ of e^{-Ht} satisfies*

$$K(t, x, y) \geq c t^{-N/2} \exp\{-a(x-y)^2/t\}$$

for all $x, y \in \mathbb{R}^N$ and $t > 0$.

Proof. Since b and $\beta > 0$ are unconstrained in Lemma 3.3.3 when $\Omega = \mathbb{R}^N$, we may put $t = T_5 \beta^2$ and $(x+y)/2 = b$. We discover that there exist $c_6 > 0$ and $c_7 > 0$ such that $|x - y| \leq c_6 t^{\frac{1}{2}}$ implies

$$K(t, x, y) \geq c_7 t^{-N/2}.$$

Now let $t > 0$ and $x, y \in \mathbb{R}^N$ be arbitrary. If we put

$$x_r = x + r(y - x)/M$$

for $0 \leq r \leq M$ then

$$|x_r - x_{r+1}| \leq \tfrac{1}{2} c_6 (t/M)^{\frac{1}{2}}$$

if and only if

$$4(y-x)^2/c_6^2 t \leq M.$$

We take M to be the smallest integer which achieves this inequality. Then

$$K(t, x, y) \geq \int K(t/M, x, y_1) K(t/M, y_1, y_2) \cdots K(t/M, y_{M-1}, y) \, dy_1 \cdots dy_{M-1}$$

where we integrate y_r over the set

$$\{y_r : |y_r - x_r| < \tfrac{1}{4} c_6 (t/M)^{\frac{1}{2}}\}.$$

This yields the bound

$$K(t, x, y) \geq \{c_7 (t/M)^{-N/2}\}^M \{k(t/M)^{N/2}\}^{M-1}$$
$$\geq c_8 t^{-N/2} k_1^M$$

which implies a lower bound of the stated type.

If we replace $\mathbb{R}^N$ by Ω in Theorem 3.3.4 the result cannot remain true even if we replace $|x - y|$ by the geodesic distance between x and y with Ω, because in this situation one knows that $K(t, x, y) \to 0$ as $x \to \partial\Omega$ or $y \to \partial\Omega$.

However, such a lower bound is not even true if one restricts x and y to lie away from $\partial\Omega$ unless one allows $b > 0$ to depend upon the geometry of Ω. The reason is that if x and y lie in different parts of a nearly disconnected region Ω, the magnitude of $K(t,x,y)$ will be very small. One does have the following result.

Theorem 3.3.5. *Let Ω be a region in $\mathbb{R}^N$ and let H be an elliptic operator on $L^2(\Omega)$ with Dirichlet boundary conditions. If A is a compact subset of Ω and $t > 0$ then there exists a continuous positive function $c(A,t)$ of t such that the heat kernel of e^{-Ht} satisfies*

$$K(t,x,y) \geqslant c(A,t)$$

for all $x, y \in A$. In particular e^{-Ht} is irreducible.

Proof. Let Ω' be a region in $\mathbb{R}^N$ with compact closure such that $A \subseteq \Omega' \subseteq \bar{\Omega}' \subseteq \Omega$ and let H' be the elliptic operator on $L^2(\Omega')$ associated with the same quadratic form as H. Then H' is uniformly elliptic because of our precise definition of ellipticity in (1.2.7) and

$$K(t,x,y) \geqslant K'(t,x,y) \geqslant 0$$

for all $x, y \in \Omega'$ and $t > 0$ by Theorem 2.1.6. It is therefore sufficient to treat the case where H is uniformly elliptic.

By the compactness of A there exists $\delta > 0$ and $D < \infty$ such that every $x, y \in A$ can be connected by a path $\gamma:[0,1] \to \Omega$ such that γ has length $\leqslant D$ and every point of γ has distance $\geqslant \delta$ from $\partial\Omega$. We now put $x_r = \gamma(r/M)$ and $t_r = tr/M$ and proceed as in the proof of Theorem 3.3.4, taking M large enough so that the conditions of Lemma 3.3.3 are satisfied. $\square$

The above theorem leads immediately to two versions of the Harnack inequality.

Corollary 3.3.6. *Let H, Ω, A be as in Theorem 3.3.5 and let $0 < a \leqslant b < \infty$. If*

$$u(x,t) = (e^{-Ht}u)(x)$$

where $0 \leqslant u \in L^1 + L^\infty$ is not identically zero, then there exists $c > 0$ such that

$$u(x,t) \geqslant c$$

for all $x \in A$ and $a \leqslant t \leqslant b$.

Proof. By expanding A if necessary we may assume that

$$\int_A u(y)\,dy \geqslant k > 0$$

Theorem 3.3.5 now implies

$$u(x,t) \geqslant \int_A K(t,x,y)u(y)\,dy$$
$$\geqslant c(A,t)k.$$

The proof is completed by putting

$$c = k\min\{c(A,t): a \leqslant t \leqslant b\}.$$

Corollary 3.3.7. *If Ω is a bounded region in $\mathbb{R}^N$ then the smallest eigenvalue E_0 of H has multiplicity one, and the corresponding eigenfunction ϕ satisfies*

$$\phi(x) \geqslant c(A) > 0$$

for any compact set $A \subseteq \Omega$ and all $x \in A$.

Proof. The first statement of the corollary follows immediately from Proposition 1.4.3. Secondly if $\phi \geqslant 0$, $\|\phi\|_2 = 1$ and $H\phi = E_0\phi$ then there exists a compact set $B \subseteq \Omega$ such that

$$\int_B \phi\,dx = c > 0.$$

Then $x \in A$ implies

$$\phi(x) = e^{E_0}\int K(1,x,y)\phi(y)\,dy$$
$$\geqslant e^{E_0}c(A \cup B, 1)c$$

by Theorem 3.3.5.

Although the above theorems provide information which will be vital in subsequent calculations, they are not very precise quantitatively. It turns out that more precise upper and lower bounds on the ground state ϕ for a region Ω depend both upon the local regularity properties of $\partial\Omega$ and upon the global geometry of Ω. These issues will be investigated in Chapter 4, but a lot remains to be done even when $H = -\Delta$ and Ω is a piecewise smooth region.

3.4 Functions of an elliptic operator

In this section we assume that Ω is a locally compact second countable metric space with metric d, and that dx is a regular Borel measure on Ω, such that the measure of every point is zero (dx is continuous). We assume that e^{-Ht} is a symmetric Markov semigroup on $L^2(\Omega)$ whose heat kernel satisfies

$$0 \leqslant K(t,x,y) \leqslant ct^{-N/2}e^{-bd^2/t} \qquad (3.4.1)$$

for all $t > 0$ and $x, y \in \Omega$, where b, c, N are positive constants. An example of the above situation is where H is a uniformly elliptic operator on $L^2(\mathbb{R}^N)$; see Corollary 3.2.8.

If f is a bounded measurable function on $[0, \infty)$ then $f(H)$ is well defined as a bounded operator on $L^2(\Omega)$. We say that a function $K_f(x, y)$ on $\Omega \times \Omega$ which is locally bounded away from the diagonal is the integral kernel of $f(H)$ if

$$\langle f(H)\phi, \psi \rangle = \int_{\Omega \times \Omega} K_f(x, y)\phi(y)\overline{\psi(x)}\,dx\,dy \qquad (3.4.2)$$

for all $\phi, \psi \in C_c(\Omega)$ with disjoint supports. It is clear from (3.4.2) that f determines K_f uniquely up to almost everywhere equivalence. By using a partition of unity, the fact that dx is continuous, and the assumed boundedness of f, it may conversely be seen that the kernel K_f determines the operator $f(H)$ uniquely. If also $K_f \in L^1_{\text{loc}}(\Omega \times \Omega)$ then (3.4.2) holds for all $\phi, \psi \in C_c(\Omega)$, but we do not make this assumption.

We shall obtain pointwise bounds on K_f for various f by two different methods. Although the first method is of rather restricted applicability, it does yield upper bounds on a variety of Green functions.

Lemma 3.4.1. *If*

$$f(s) = \int_0^\infty e^{-st}\rho(t)\,dt$$

where $\rho \in L^1(0, \infty)$ then

$$|K_f| \leq c_1 \|\rho\|_1 d^{-N}.$$

Proof. By combining the formula

$$K_f(x, y) = \int_0^\infty K(t, x, y)\rho(t)\,dt$$

with (3.4.1) we obtain

$$|K_f| \leq \int_0^\infty ct^{-N/2}e^{-bd^2/t}|\rho(t)|\,dt$$

$$\leq \int_0^\infty c_2 t^{-N/2}(bd^2/t)^{-N/2}|\rho(t)|\,dt$$

$$= c_1 d^{-N}\|\rho\|_1.$$

Lemma 3.4.2. *If*

$$f(s) = (s + \lambda)^{-\alpha}$$

where $\alpha > N/2$ and $\lambda > 0$ then

$$|K_f| \leq c_\beta e^{-\beta d}$$

for all $0 < \beta < 2(b\lambda)^{\frac{1}{2}}$.

Proof. By combining the formula

$$(s+\lambda)^{-\alpha} = \Gamma(\alpha)^{-1} \int_0^\infty t^{\alpha-1} e^{-t(s+\lambda)} \, dt$$

with (3.4.1) we obtain

$$|K_f| \leq \int_0^\infty c t^{-N/2} e^{-bd^2/t} \Gamma(\alpha)^{-1} t^{\alpha-1} e^{-t\lambda} \, dt$$

$$= a \int_0^\infty t^{\alpha-N/2-1} e^{-\delta t} \exp\{(\delta-\lambda)t - bd^2/t\} \, dt \qquad (3.4.3)$$

for any $0 < \delta < \lambda$. But

$$0 \leq \exp\{(\delta-\lambda)t - bd^2/t\} \leq \exp\{-2(\lambda-\delta)^{\frac{1}{2}} b^{\frac{1}{2}} d\}$$

for all $0 < t < \infty$, so putting $\beta = 2(\lambda-\delta)^{\frac{1}{2}} b^{\frac{1}{2}}$ yields

$$|K_f| \leq a e^{-\beta d} \int_0^\infty t^{\alpha-N/2-1} e^{-\delta t} \, dt$$

as required.

Lemma 3.4.3. *If*

$$f(s) = (s+\lambda)^{-\alpha}$$

where $\alpha < N/2$ and $\lambda > 0$ then

$$|K_f| \leq c_\beta d^{2\alpha-N} e^{-\beta d}$$

for all $0 < \beta < 2(b\lambda)^{\frac{1}{2}}$. If $\lambda = 0$ the same holds with $\beta = 0$.

Proof. We deduce from (3.4.3) that

$$|K_f| \leq a d^{2\alpha-N} \int_0^\infty s^{\alpha-N/2-1} \exp(-b/s - \lambda d^2 s) \, ds$$

$$= a d^{2\alpha-N} \int_0^\infty s^{\alpha-N/2-1} e^{-\delta/s} \exp\{-(b-\delta)/s - \lambda d^2 s\} \, ds$$

for any $0 < \delta < b$. But

$$0 \leq \exp\{-(b-\delta)/s - \lambda d^2 s\} \leq \exp\{-2(b-\delta)^{\frac{1}{2}} \lambda^{\frac{1}{2}} d\}$$

for all $0 < s < \infty$, so putting $\beta = 2(b-\delta)^{\frac{1}{2}} \lambda^{\frac{1}{2}}$ yields

$$|K_f| \leq a d^{2\alpha-N} e^{-\beta d} \int_0^\infty s^{\alpha-N/2-1} e^{-\delta/s} \, ds$$

as required.

Note 3.4.4. The sharpness of these bounds can be seen by comparing them with the exact expressions for $H = -\Delta$ on $\mathbb{R}^N$, computed by Fourier transform techniques. For example if $\alpha = 1$ and $N = 3$ then Lemma 3.4.3 yields

$$|K_f| \leq c_\delta |x-y|^{-1} \exp\{-(1-\delta)\lambda^{\frac{1}{2}}|x-y|\} \qquad (3.4.4)$$

for all $\delta > 0$, whereas the exact expression is

$$K_f = (4\pi|x-y|)^{-1} \exp(-\lambda^{\frac{1}{2}}|x-y|).$$

However if $N = 5$ then one finds that one cannot set $\beta = 2(b\lambda)^{\frac{1}{2}}$ in Lemma 3.4.3.

Lemma 3.4.5. *If*

$$|f(s)| \leq a(1+s)^{-\alpha}$$

for all $0 < s < \infty$ and some $a > 0$, $\alpha > N/2$, then K_f is a bounded integral kernel.

Proof. We put

$$f(s) = (1+s)^{-\alpha}g(s)$$

where g is a bounded function on $(0, \infty)$, so $g(H)$ is a bounded operator on $L^2(\Omega)$. Lemma 3.4.2 implies that $(1+H)^{-\alpha}$ is a bounded operator from L^1 to L^∞, so complex interpolation implies that $(1+H)^{-\alpha/2}$ is bounded from L^1 to L^2 and also from L^2 to L^∞. Therefore

$$f(H) = (1+H)^{-\alpha/2}g(H)(1+H)^{-\alpha/2}$$

is bounded from L^1 to L^∞ and has a bounded integral kernel.

We now turn to our second method. We obtain bounds on the heat kernel $K(z, x, y)$ for complex times $\operatorname{Re} z > 0$, and use them to bound K_f for suitable f by Fourier transform techniques. This method is much more powerful than the previous one.

Lemma 3.4.6. *The heat kernel satisfies*

$$|K(z,x,y)| \leq c_1 (\operatorname{Re} z)^{-N/2}$$

for all $\operatorname{Re} z > 0$ and $x, y \in \Omega$.

Proof. Putting $z = t + is$ we have

$$e^{-Hz} = e^{-Ht/2}e^{iHs}e^{-Ht/2}$$

so

$$\|e^{-Hz}\|_{\infty,1} \leq \|e^{-Ht/2}\|_{\infty,2}\|e^{-Ht/2}\|_{2,1}$$
$$= \|e^{-Ht/2}\|_{\infty,2}^2.$$

Functions of an elliptic operator

One sees by interpolation that
$$\|e^{-Ht/2}\|_{\infty,2}^2 \leq \|e^{-Ht/2}\|_{\infty,1}$$
$$\leq c(t/2)^{-N/2}$$
so the lemma follows.

Note 3.4.7. If $H = -\Delta$ on $L^2(\mathbb{R}^N)$ then
$$K(z,x,y) = (4\pi z)^{-N/2} e^{-|x-y|^2/4z}$$
so
$$|K(z,x,y)| \leq (4\pi|z|)^{-N/2}$$
for all $\operatorname{Re} z > 0$. However if $H = -d^2/dx^2$ on $L^2(0,\pi)$ with periodic boundary conditions then
$$K(z,x,y) = K(z + 2\pi i, x, y)$$
for all $\operatorname{Re} z > 0$, so no bound of the form
$$|K(z,x,y)| \leq c|z|^{-N/2}$$
is possible for this example.

Theorem 3.4.8. *The complex time heat kernel satisfies*
$$|K(z,\omega_1,\omega_2)| \leq c_\varepsilon(\operatorname{Re} z)^{-N/2} \exp\{-\operatorname{Re}(bd^2/(1+\varepsilon)z)\}$$
for all $\operatorname{Re} z > 0$ and $\omega_i \in \Omega$ and $\varepsilon > 0$.

Proof. Define the analytic function g in the sector
$$D = \{z: 0 \leq \arg z \leq \gamma\}$$
where $0 < \gamma < \pi/2$, by
$$g(z) = z^{-N/2} K(z^{-1}, \omega_1, \omega_2) \exp\{bd^2 e^{i(\pi/2-\gamma)} z/\sin\gamma\}$$
If $z = re^{i\theta}$ and $\theta = 0$ then
$$|g(z)| \leq r^{-N/2} cr^{N/2} e^{-bd^2 r} \exp\{(bd^2 r/\sin\gamma)\operatorname{Re} e^{i(\pi/2-\gamma)}\}$$
$$= c.$$
If $z = re^{i\theta}$ and $\theta = \gamma$ then Lemma 3.4.6 implies
$$|g(z)| \leq r^{-N/2} c_1(\operatorname{Re} z^{-1})^{-N/2}$$
$$= c_1(\sec\gamma)^{N/2}.$$
If $z = re^{i\theta}$ and $0 \leq \theta \leq \gamma$ then
$$|g(z)| \leq r^{-N/2} c_1(\operatorname{Re} z^{-1})^{-N/2} \exp\{(bd^2 r/\sin\gamma)\operatorname{Re} e^{i(\pi/2-\gamma+\theta)}\}$$
$$\leq c_1(\sec\gamma)^{N/2} \exp\{db^2 r/\sin\gamma\}.$$

The Phragmén–Lindelöf theorem now implies that
$$|g(z)| \leq c_2(\sec\gamma)^{N/2}$$
for all $z \in D$. Now
$$K(z, \omega_1, \omega_2) = z^{-N/2} g(z^{-1}) \exp\{-bd^2 e^{i(\pi/2 - \gamma)}/z \sin\gamma\}.$$
Therefore $-\gamma \leq \operatorname{Arg} z \leq 0$ implies
$$|K(z, \omega_1, \omega_2)| \leq r^{-N/2} c_2(\sec\gamma)^{N/2} \exp\{-\operatorname{Re} bd^2 e^{i(\pi/2 - \gamma)}/z \sin\gamma\}$$
$$= c_2(\operatorname{Re} z)^{-N/2}(\cos\theta/\cos\gamma)^{N/2} \exp\{-bd^2 \sin(\gamma + \theta)/r \sin\gamma\}.$$
Replacing z by $\bar{z}$ we deduce that
$$|K(z, \omega_1, \omega_2)| \leq c_2(\operatorname{Re} z)^{-N/2}(\cos\theta/\cos\gamma)^{N/2} \exp\{-bd^2 \sin(\gamma - |\theta|)/r \sin\gamma\}$$
provided
$$|\theta| = |\arg z| \leq \gamma < \pi/2.$$
This inequality is satisfied for all $\operatorname{Re} z > 0$ and if $0 < \varepsilon < 1$ we put
$$\gamma = (\pi/2)(1 - \varepsilon) + \varepsilon|\theta|.$$
Using the bound
$$\sin(\varepsilon\phi) \geq \varepsilon \sin\phi$$
valid for all $0 \leq \phi \leq \pi/2$ and $0 < \varepsilon < 1$ we deduce that
$$|K(z, \omega_1, \omega_2)| \leq c_2(\varepsilon \operatorname{Re} z)^{-N/2} \exp\{-bd^2(1 - \varepsilon)\cos\theta/r \sin\gamma\}$$
$$\leq c_2(\varepsilon \operatorname{Re} z)^{-N/2} \exp\{-bd^2(1 - \varepsilon)\operatorname{Re}(z^{-1})\}$$
from which the theorem is immediate.

There are very many applications of the above bounds. The following two are representative, but the method used is more important than the particular results obtained.

Theorem 3.4.9. *If $f \in C_c^1(\mathbb{R})$ and $f'(u)$ is piecewise continuously differentiable then*
$$|K_f| \leq c_\beta(1 + d)^{-\beta}$$
for all $0 \leq \beta < 1$.

Proof. We observe that
$$f(u) = e^{-u} \int_{-\infty}^{\infty} e^{-ius} g(s) \, ds$$
where
$$|g(s)| \leq c(1 + s^2)^{-1}.$$
Therefore
$$f(H) = \int_{-\infty}^{\infty} e^{-H(1 + is)} g(s) \, ds$$

and

$$|K_f| \leqslant \int_{-\infty}^{\infty} c_\varepsilon \exp\{-bd^2/(1+\varepsilon)(1+s^2)\}c(1+s^2)^{-1}\,ds$$

$$\leqslant a_\beta d^{-\beta} \int_{-\infty}^{\infty} (1+s^2)^{\beta/2-1}\,ds$$

provided $0 \leqslant \beta < 1$. The theorem follows by combining this with the same bound for $\beta = 0$.

Theorem 3.4.10. *If f lies in the Schwartz space $\mathscr{S}$ then*

$$|K_f| \leqslant c_\beta (1+d)^{-\beta}$$

for all $\beta \geqslant 0$.

Proof. Since $H \geqslant 0$ we may assume that $f(u) = 0$ for $u \leqslant -\frac{1}{2}$, and can then write

$$f(u) = (1+u)^{-n} \int_{-\infty}^{\infty} e^{-ius} g_n(s)\,ds$$

where $g_n \in \mathscr{S}$ for all $n \geqslant 0$. Therefore

$$f(u) = \Gamma(n)^{-1} \int_D e^{-u(t+is)} t^{n-1} e^{-t} g_n(s)\,ds\,dt$$

where

$$D = \{(t,s): t > 0 \quad \text{and} \quad s \in \mathbb{R}\}.$$

Applying Theorem 3.4.8 and assuming that $\beta \geqslant 0$ we obtain

$$|K_f| \leqslant c \int_D t^{n-1-N/2} |g_n(s)| \exp\{-t - bd^2 t/(1+\varepsilon)(s^2+t^2)\}\,ds\,dt$$

$$\leqslant c_4 d^{-\beta} \int_D t^{n-1-N/2-\beta/2} |g_n(s)| e^{-t} (s^2+t^2)^{\beta/2}\,ds\,dt$$

where the integral is finite provided n is large enough. The theorem follows by combining this with the same bound for $\beta = 0$.

Notes

Section 3.2 The validity of Gaussian upper and lower bounds of the form (3.1.3) for uniformly elliptic operators on $L^2(\mathbb{R}^N)$ with measurable coefficients was proved by Aronson (1968); see also Porper and Eidel'man (1984) for references to the Soviet literature on this problem. Theorem 3.2.7, due to

Davies (1987B), is sharper in that it replaces $b_2|x-y|^2$ in the upper bound by $d(x,y)^2/(4+\varepsilon)$. We mention that Varopoulos (1985D) and Carlen, Kusuoka and Stroock (1987) have obtained Gaussian upper bounds for large times even when the operator H is non-local.

Example 3.2.11 is a mere taste of the difficult and deep results on hypoelliptic operators of second order, which have been studied by Fefferman and Sánchez-Calle (1986), Jerison and Sánchez-Calle (1986), Kusuoka and Stroock (1987; 1988), Léandre (1987A; 1987B), Nagel, Stein and Waigner (1985), Sánchez-Calle (1984), Varopoulos (1985–6) and others. Davies (1988B) showed that the method of Theorem 3.2.7 can be extended to obtain sharper upper bounds of the heat kernel than previously known.

For the case where $H = -\Delta$ on a region $\Omega \subseteq \mathbb{R}^N$ one would expect to be able to obtain far more detailed results than we have given here. For such bounds see van den Berg (1987A; 1987B) and references there.

Section 3.3 Gaussian lower bounds on heat kernels of uniformly elliptic operators on $\mathbb{R}^N$ have been known for a long time; see Aronson (1968) and Porper and Eidel'man (1984). However, our approach is a minor adaptation of that of Fabes and Stroock (1986), who carried through to completion ideas originating from Nash (1958). For Moser's original proof of the parabolic Harnack inequality of Corollary 3.3.6 see Moser (1964; 1967). For extensions of the elliptic Harnack inequality of Corollary 3.3.7 see Moser (1961) and Gilbarg and Trudinger (1977).

Section 3.4 This section is taken from Davies (1988A), except for the improved version of Theorem 3.4.8, which was communicated to the author by B. Simon. For the Laplace operator on a manifold an approach based upon the wave equation may be found in Cheeger, Gromov and Taylor (1982), and this can probably be extended to elliptic operators with smooth coefficients.

4
Boundary behaviour

4.1 Introduction

In this chapter we shall be concerned with obtaining sharper bounds on the heat kernel $K(t, x, y)$ of an elliptic operator H on $L^2(\Omega)$ near the boundary $\partial\Omega$ of Ω, or near infinity if $\Omega = \mathbb{R}^N$. The problem with earlier results such as Example 2.1.8 is that they do not draw attention to the fact that if one has Dirichlet boundary conditions one expects that $K(t, x, y) \to 0$ as $x \to \partial\Omega$ or $y \to \partial\Omega$. Quantitatively the problem is to obtain bounds on the rate of convergence.

The following simple examples illuminate much of our subsequent discussion.

Example 4.1.1. Let K be the heat kernel of $H = -\Delta$ on $L^2(0, \infty)$ subject to Dirichlet boundary conditions. The reflection principle states that the heat kernel is

$$K(t, x, y) = (4\pi t)^{-\frac{1}{2}} e^{-(x-y)^2/4t} - (4\pi t)^{-\frac{1}{2}} e^{(x+y)^2/4t}$$
$$= (4\pi t)^{-\frac{1}{2}} e^{-(x-y)^2/4t} [1 - e^{-xy/t}]$$

for all positive x, y and t. Therefore there exist $c_i > 0$ such that

$$c_1 t^{-\frac{1}{2}} e^{-(x-y)^2/4t}(1 \wedge xy/t) \leqslant K(t, x, y)$$
$$\leqslant c_2 t^{-\frac{1}{2}} e^{-(x-y)^2/4t}(1 \wedge xy/t).$$

In particular

$$K(t, x, x) \sim t^{-\frac{1}{2}}(1 \wedge x^2/t)$$

for all positive x and t, so one has different asymptotic behaviour as $x^2/t \to 0$ and as $x^2/t \to +\infty$. Although there is a global bound of the form

$$0 \leqslant K(t, x, y) \leqslant ct^{-\frac{1}{2}}$$

a bound of the form

$$0 \leqslant K(t, x, y) \leqslant ct^{-\beta} xy$$

holds for $0 < t \leqslant 1$ if and only if $\beta \geqslant \frac{3}{2}$.

Example 4.1.2. Let K_N be the heat kernel of $H = -\Delta$ on $L^2((0,\infty)^N)$ subject to Dirichlet boundary conditions. Then

$$K_N(t,x,y) = \prod_{i=1}^{N} K(t,x_i,y_i)$$

where K is the heat kernel of the previous example. In particular if $N = 3$ then

$$K(t,x,x) \sim t^{-\frac{3}{2}} \min\left\{1, \frac{x_1^2}{t}, \frac{x_2^2}{t}, \frac{x_3^2}{t}, \frac{x_1^2 x_2^2}{t^2}, \frac{x_2^2 x_3^2}{t^2}, \frac{x_3^2 x_1^2}{t^2}, \frac{x_1^2 x_2^2 x_3^2}{t^3}\right\}.$$

The point is that even for this simple region the heat kernel has different asymptotic forms on the diagonal $x = y$ in each of eight subregions of $(0,\infty)^3$.

Example 4.1.3. Let $H = -\Delta$ on $\Omega = \{x \in \mathbb{R}^N : x_1 > 0\}$ subject to Dirichlet boundary conditions, where $N \geqslant 3$. The reflection principle implies that the Green function is of the form

$$G(x,y) = c_N|x-y|^{2-N} - c_N|x-z|^{2-N}$$

where $z = (-y_1, y_2, y_3, \ldots, y_N)$. Now

$$|x-z|^2 = |x-y|^2 + 4x_1 y_1$$
$$= |x-y|^2(1+\gamma)$$

where $\gamma > 0$. Therefore

$$G(x,y) = c_N|x-y|^{2-N}\{1 - (1+\gamma)^{1-N/2}\}$$
$$\sim |x-y|^{2-N}(1 \wedge \gamma)$$

for all $x, y \in \Omega$. That is there exists $a_N > 0$ such that

$$a_N^{-1}\left(\frac{1}{|x-y|^{N-2}} \wedge \frac{x_1 y_1}{|x-y|^N}\right) \leqslant G(x,y)$$
$$\leqslant a_N\left(\frac{1}{|x-y|^{N-2}} \wedge \frac{x_1 y_1}{|x-y|^N}\right)$$

for all $x, y \in \Omega$ provided $N \geqslant 3$.

We now return to the general theory. If H has compact resolvent, there exists a complete orthonormal set $\{\phi_n\}_{n=0}^{\infty}$ such that $H\phi_n = E_n\phi_n$ where $E_n \to \infty$ as $n \to \infty$. One then has

$$K(t,x,y) = \sum_{n=0}^{\infty} \exp(-E_n t)\phi_n(x)\phi_n(y)$$

and in particular

$$K(t,x,x) = \sum_{n=0}^{\infty} \exp(-E_n t)|\phi_n(x)|^2.$$

One therefore sees that K cannot vanish at the boundary faster than the eigenfunctions ϕ_n (which need not vanish at the same rate for different n). We shall see in a surprisingly wide range of cases that the behaviour of K near $\partial\Omega$ is controlled by that of the ground state ϕ_0. This then reduces the problem to one about which a great deal can be said. Note, however, that if $\partial\Omega$ is piecewise smooth, the rate at which $\phi_0(x)$ vanishes as $x \to x_0 \in \partial\Omega$ depends strongly both upon the coefficients of H and upon the geometry of $\partial\Omega$ around the point x_0; see Section 4.6 for details.

4.2 Transference to weighted L^2 spaces

In this section we introduce a new method for studying elliptic operators or Schrödinger operators which is based upon the calculation in Theorem 1.5.12. Throughout this section we assume that H is defined as a quadratic form on $C_c^1(\Omega)$ by

$$Q(f) = \int \left\{ \sum a_{ij} \frac{\partial f}{\partial x_i} \frac{\partial \bar{f}}{\partial x_j} + V|f|^2 \right\} dx$$

where a is a positive C^1 matrix-valued function and $V \in L^1_{\text{loc}}(\Omega)$. Several technical variations upon the conditions above and those of the theorem below are possible.

Theorem 4.2.1. *Suppose that there exists a positive C^2 function ϕ on Ω such that*

$$-\sum \frac{\partial}{\partial x_i}\left(a_{ij} \frac{\partial \phi}{\partial x_j}\right) + V\phi \geq 0 \qquad (4.2.1)$$

on Ω. Then the form Q is non-negative and closable on $C_c^1(\Omega)$. The corresponding self-adjoint operator H on $L^2(\Omega, dx)$ generates a positivity-preserving semigroup e^{-Ht}.

Proof. If we put

$$X = \phi^{-1} \sum \frac{\partial}{\partial x_i}\left(a_{ij} \frac{\partial \phi}{\partial x_j}\right)$$

then X is continuous and $V \geq X$ by (4.2.1). If $f \in C_c^1(\Omega)$ and we put $g = \phi^{-1} f$ then

$$\begin{aligned}
Q(f) &= \int \left\{ \sum a_{ij} \frac{\partial(\phi g)}{\partial x_i} \frac{\partial(\phi \bar{g})}{\partial x_j} + V\phi^2 |g|^2 \right\} dx \\
&= \int \left\{ \sum a_{ij} \left(|g|^2 \frac{\partial \phi}{\partial x_i} \frac{\partial \phi}{\partial x_j} + \phi \bar{g} \frac{\partial g}{\partial x_i} \frac{\partial \phi}{\partial x_j} \right. \right. \\
&\qquad \left. \left. + g\phi \frac{\partial \phi}{\partial x_i} \frac{\partial \bar{g}}{\partial x_j} + \phi^2 \frac{\partial g}{\partial x_i} \frac{\partial \bar{g}}{\partial x_j} \right) + V\phi^2 |g|^2 \right\} dx
\end{aligned}$$

$$= \int \left\{ \sum a_{ij} \frac{\partial \phi}{\partial x_i} \frac{\partial (\phi|g|^2)}{\partial x_j} \right.$$
$$\left. + \sum \phi^2 a_{ij} \frac{\partial g}{\partial x_i} \frac{\partial \bar{g}}{\partial x_j} + V\phi^2|g|^2 \right\} dx$$
$$= \int \left\{ \sum a_{ij} \frac{\partial g}{\partial x_i} \frac{\partial \bar{g}}{\partial x_j} + (V - X)|g|^2 \right\} \phi^2 \, dx.$$

This immediately establishes that the form Q is non-negative. More importantly if we define the unitary operator U_ϕ from $L^2(\Omega, \phi^2 \, dx)$ onto $L^2(\Omega, dx)$ by $U_\phi f = \phi f$ then U_ϕ maps $C_c^1(\Omega)$ onto $C_c^1(\Omega)$. The new form

$$Q_\phi(g) = Q(U_\phi g)$$

defined on the dense subspace $C_c^1(\Omega)$ of $L^2(\Omega, \phi^2 \, dx)$, is given by

$$Q_\phi(g) = \int \left\{ \sum a_{ij} \frac{\partial g}{\partial x_i} \frac{\partial \bar{g}}{\partial x_j} + (V - X)|g|^2 \right\} \phi^2 \, dx.$$

The next stage in the argument is to apply Theorem 1.8.1, which can be done once one checks that the replacement of dx by $\phi^2 \, dx$ in its proof and the proof of all the other results upon which it depends does not matter. This allows us to construct a self-adjoint operator $H_\phi \geq 0$ on $L^2(\Omega, \phi^2 \, dx)$ such that $\exp(-H_\phi t)$ is a symmetric Markov semigroup. The theorem now follows upon observing that $U^{\pm 1}$ are both unitary and positivity-preserving.

Although the operator U_ϕ takes $L^p(\Omega, \phi^2 \, dx)$ isometrically one–one onto $L^p(\Omega, dx)$ when $p = 2$, there is no such relationship between the L^p spaces for other values of p in general. In particularly the ultracontractivity of e^{-Ht} and of $\exp(-H_\phi t)$ are quite independent properties.

Lemma 4.2.2. *The semigroup $\exp(-H_\phi t)$ is ultracontractive if and only if there are constants $c_t > 0$ for all $0 < t < \infty$ such that the heat kernel $K(t, x, y)$ of e^{-Ht} satisfies*

$$0 \leq K(t, x, y) \leq c_t \phi(x) \phi(y). \tag{4.2.2}$$

Proof. If $f \in L^2(\Omega, \phi^2 \, dx)$ then

$$\exp(-H_\phi t) f(x) = \phi(x)^{-1} e^{-Ht}(\phi f)(x)$$
$$= \int \phi(x)^{-1} K(t, x, y) \phi(y) f(y) \, dy$$

so the heat kernel of $\exp(-H_\phi t)$ is

$$K_\phi(t, x, y) = \phi(x)^{-1} K(t, x, y) \phi(y)^{-1}.$$

The lemma follows immediately.

Theorem 4.2.3. *Suppose that the positive C^2 function ϕ satisfies (4.2.1) and that*
$$\|\exp(-H_\phi t)\|_{\infty,2} \leq c_t$$
for all $0 < t < \infty$. Then if $f \in \mathrm{Dom}(H) \subseteq L^2(\Omega, dx)$ satisfies $Hf = Ef$, one has
$$|f(x)| \leq \phi(x) \|f\|_2 \inf\{e^{Et}c_t : 0 < t < \infty\}.$$

Proof. If we put $g = \phi^{-1}f$ then $H_\phi g = Eg$ and
$$\begin{aligned}\|g\|_\infty &= e^{Et}\|\exp(-H_\phi t)g\|_\infty \\ &\leq e^{Et}c_t\|g\|_2 \\ &= e^{Et}c_t\|f\|_2.\end{aligned}$$

The above theorems have not assumed any boundedness or L^p properties of the function ϕ which satisfies (4.2.1). However, if $\phi \in L^2(\Omega, dx)$ then (4.2.2) implies that e^{-Ht} has a Hilbert–Schmidt kernel, and hence that H has compact resolvent and discrete spectrum.

We next describe a technical variation of the above theorem. We suppose that Ω is a locally compact, second countable, Hausdorff space and that dx is a Borel measure on Ω. We suppose that $H = H^*$ is a semibounded operator on $L^2(\Omega, dx)$ and that e^{-Ht} is an irreducible positivity-preserving semigroup. We assume that the bottom of the spectrum E_0 is an eigenvalue. By Proposition 1.4.3, E_0 has multiplicity one and the corresponding eigenfunction ϕ_0, normalised by $\|\phi_0\|_2 = 1$, is positive almost everywhere.

We now define the unitary operator U from $L^2(\Omega, \phi_0^2 dx)$ to $L^2(\Omega, dx)$ by $Uf = \phi_0 f$ and define $\tilde{H}$ on $L^2(\Omega, \phi_0^2 dx)$ by
$$\tilde{H} = U^{-1}(H - E_0)U$$
so that
$$\mathrm{Dom}(\tilde{H}^\alpha) = U^{-1}\mathrm{Dom}(H^\alpha)$$
for any $\alpha \geq 0$. It is evident from the definitions that $e^{-\tilde{H}t}$ is positivity-preserving and that
$$e^{-\tilde{H}t}1 = \phi_0^{-1}\{e^{(E_0 - H)t}\phi_0\} = 1$$
for all $t \geq 0$. Therefore $-1 \leq f \leq 1$ implies
$$-1 \leq e^{-\tilde{H}t}f \leq 1$$
for all $t \geq 0$ and $e^{-\tilde{H}t}$ is an irreducible symmetric Markov semigroup on $L^2(\Omega, \phi_0^2 dx)$.

Theorem 4.2.4. *If $e^{-\tilde{H}t}$ is ultracontractive with*
$$\|e^{-\tilde{H}t}\|_{\infty,2} \leq c_t$$
for all $0 < t < \infty$, then there exists a complete orthonormal set $\{\phi_n\}_{n=0}^\infty$ in

$L^2(\Omega, dx)$ such that $H\phi_n = E_n\phi_n$, and there exist constants c_n such that

$$|\phi_n(x)| \leq c_n \phi_0(x) \qquad (4.2.3)$$

for all $n \geq 0$ and $x \in \Omega$.

Proof. The kernel $\tilde{K}(t, x, y)$ of $e^{-\tilde{H}t}$ is given by

$$\tilde{K}(t, x, y) = e^{E_0 t} \phi_0(x)^{-1} K(t, x, y) \phi_0(y)^{-1}$$

and the ultracontractivity assumption implies that

$$0 \leq \tilde{K}(t, x, y) \leq c_t^2.$$

Since $(\Omega, \phi_0^2 dx)$ is a probability space, it follows that $\tilde{K}$ is a Hilbert–Schmidt kernel. Therefore $\tilde{H}$ has a complete orthonormal set of eigenfunctions $\{\tilde{\phi}_n\}_{n=0}^{\infty}$ which satisfy

$$|\tilde{\phi}_n(x)| \leq c_n \equiv \inf\{c_t e^{(E_n - E_0)t} : 0 < t < \infty\}$$

as in Theorem 4.2.3. The theorem follows by putting

$$\phi_n = U(\tilde{\phi}_n) = \phi \cdot \tilde{\phi}_n.$$

It is a remarkable fact that one can obtain lower bounds on the heat kernel and Green function from the ultracontractivity assumptions. It will be seen that the constants below depend upon the magnitude of the positive gap $(E_1 - E_0)$ in the spectrum.

Theorem 4.2.5. If e^{-Ht} is ultracontractive then for any $\varepsilon > 0$ there exists T such that $t \geq T$ implies

$$(1 - \varepsilon) e^{-E_0 t} \phi_0(x) \phi_0(y) \leq K(t, x, y) \leq (1 + \varepsilon) e^{-E_0 t} \phi_0(x) \phi_0(y)$$

for all $x, y \in \Omega$. If $G(\lambda, x, y)$ is the kernel of $(H + \lambda)^{-1}$ then there exists $a_\lambda > 0$ such that

$$G(\lambda, x, y) \geq a_\lambda \phi_0(x) \phi_0(y)$$

for all $\lambda > -E_0$ and $x, y \in \Omega$.

Proof. The first bound of the theorem may be recast in the form

$$|\tilde{K}(t, x, y) - 1| < \varepsilon$$

if $t \geq T$ and $x, y \in \Omega$. Theorem 2.1.4 yields

$$\tilde{K}(t, x, y) = \sum_{n=0}^{\infty} \exp\{-(E_n - E_0)t\} \tilde{\phi}_n(x) \tilde{\phi}_n(y)$$

with $\tilde{\phi}_0 = 1$ and

$$\|\tilde{\phi}\|_\infty \leq c_{t/3} \exp\{(E_n - E_0)t/3\}$$

for all $x \in \Omega$. Therefore

$$|\tilde{K}(t,x,y) - 1| \leq c_{t/3}^2 \sum_{n=1}^{\infty} \exp\{-(E_n - E_0)t/3\}$$

which converges to zero as $t \to \infty$.

Secondly, putting $\varepsilon = \frac{1}{2}$,

$$G(\lambda, x, y) = \int_0^{\infty} e^{-\lambda t} K(t, x, y)\, dt$$

$$\geq \frac{1}{2} \int_T^{\infty} e^{-(\lambda + E_0)t} \phi_0(x)\phi_0(y)\, dt$$

$$= \frac{e^{-(\lambda + E_0)T}}{2(\lambda + E_0)} \phi_0(x)\phi_0(y)$$

for all $x, y \in \Omega$ and $\lambda > -E_0$.

The above theorems describe some consequences of the ultracontractivity of $\exp(-H_\phi t)$ or $e^{-\tilde{H}t}$, but do not indicate when this property holds. In the next section we show that for the harmonic oscillator Hamiltonian H, $e^{-\tilde{H}t}$ is hypercontractive but not ultracontractive. Later sections give a variety of more positive results.

4.3 The harmonic oscillator

In this section we treat the particular example

$$Hf = \tfrac{1}{2}(-d^2 f/dx^2 + x^2 f - f) \tag{4.3.1}$$

with domain $\mathcal{S}$ in $L^2(\mathbb{R})$. It is well known that this operator has spectrum $\{0, 1, 2, 3, \ldots\}$ and that the corresponding eigenvectors are

$$\phi_0(x) = \pi^{-\frac{1}{4}} e^{-x^2/2}$$

$$\phi_n(x) = H_n(x) e^{-x^2/2}$$

where $H_n(x)$ is a polynomial of degree n, actually a multiple of the Hermite polynomial. We quote Mehler's formula for the heat kernel $K(t, x, y)$ of e^{-Ht}.

Proposition 4.3.1. *We have*

$$K(t, x, y) = \{\pi(1 - e^{-2t})\}^{-\frac{1}{2}} \exp\left\{\frac{4xye^{-t} - (x^2 + y^2)(1 + e^{-2t})}{2(1 - e^{-2t})}\right\}$$

for all $t > 0$ and $x, y \in \mathbb{R}$.

Theorem 4.3.2. *The semigroup e^{-Ht} is ultracontractive on $L^2(\mathbb{R}, dx)$ but the semigroup $e^{-\tilde{H}t}$ is not ultracontractive on $L^2(\mathbb{R}, \phi_0^2\, dx)$.*

Proof. Since
$$2e^{-t} \leq 1 + e^{-2t}$$
we find that
$$0 \leq K(t,x,y) \leq \{\pi(1-e^{-2t})\}^{-\frac{1}{2}} \exp\{-(x-y)^2/(e^t - e^{-t})\}$$
$$\leq \{\pi(1-e^{-2t})\}^{-\frac{1}{2}}.$$
Therefore e^{-Ht} is ultracontractive. However
$$\tilde{K}(t,x,y) = \phi_0(x)^{-1} K(t,x,y) \phi_0(y)^{-1}$$
$$= (1-e^{-2t})^{-\frac{1}{2}} \exp\left\{\frac{2xye^{-t} - (x^2+y^2)e^{-2t}}{1-e^{-2t}}\right\} \quad (4.3.2)$$
which is clearly not a bounded function of x,y for any $t > 0$.

We could alternatively see that e^{-Ht} is not ultracontractive by noting that $\phi_n(x)/\phi_0(x)$ is the polynomial $\pi^{\frac{1}{4}} H_n(x)$, which is not bounded, so (4.2.3) cannot be valid.

We shall prove that $e^{-\tilde{H}t}$ is hypercontractive by means of the following lemma.

Lemma 4.3.3. *If the heat kernel $K(t,x,y)$ of a symmetric Markov semigroup e^{-Lt} on $L^2(\Omega, dx)$ satisfies*
$$0 \leq K(t,x,y) \leq c_t \phi_t(x) \psi_t(y)$$
for some $t > 0$ where $\psi_t \in L^2$ and $\phi_t \in L^{p(t)}$, then e^{-Lt} is hypercontractive.

Proof. We have
$$|e^{-Lt} f(x)| \leq |c_t \phi_t(x) \langle \psi_t, f \rangle|$$
$$\leq c_t \phi_t(x) \|\psi_t\|_2 \|f\|_2$$
so
$$\|e^{-Lt} f\|_{p(t)} \leq c_t \|\phi_t\|_{p(t)} \|\psi_t\|_2 \|f\|_2$$
as required.

For a sharp version of the following theorem see the notes.

Theorem 4.3.4. *If H is defined by (4.3.1) and $2 < p < \infty$ then $e^{-\tilde{H}t}$ is bounded from $L^2(\mathbb{R}, \phi_0^2 dx)$ to $L^p(\mathbb{R}, \phi_0^2 dx)$ provided*
$$e^t > (p-1)^{\frac{1}{2}}. \quad (4.3.3)$$

Proof. By Proposition 4.3.1 we have
$$0 \leq K(t,x,y) \leq c_t \exp\{L(t,x,y)\}$$

where
$$L(t,x,y) = \frac{2e^{-t}(\beta^{-1}x^2 + \beta y^2) - (x^2+y^2)(1+e^{-2t})}{2(1-e^{-2t})}$$
$$= -x^2 \frac{1 - 2e^{-t}\beta^{-1} + e^{-2t}}{2(1-e^{-2t})} - y^2 \frac{1 - 2e^{-t}\beta + e^{-2t}}{2(1-e^{-2t})}.$$

By choosing β appropriately we see that for any $0 < \lambda < (1-e^{-2t})/(1+e^{-2t})$ there exists $\delta > 0$ such that
$$0 \leq K(t,x,y) \leq c_t e^{-\lambda x^2/2} e^{-\delta y^2/2}.$$
Hence
$$0 \leq \tilde{K}(t,x,y) \leq \pi^{\frac{1}{2}} c_t e^{(1-\lambda)x^2/2} e^{(1-\delta)y^2/2}$$
$$= \pi^{\frac{1}{2}} c_t \phi(x)\psi(y).$$

Now calculating with respect to the measure $\phi_0^2 \, dx$ we have
$$\|\psi\|_2^2 = \int_{-\infty}^{\infty} e^{(1-\delta)y^2} \pi^{-\frac{1}{2}} e^{-y^2} \, dy < \infty$$
and
$$\|\phi\|_p^p = \int_{-\infty}^{\infty} e^{p(1-\lambda)x^2/2} \pi^{-\frac{1}{2}} e^{-x^2} \, dx.$$

This is finite provided
$$p(1-\lambda)/2 < 1$$
or
$$1 - 2/p < \lambda.$$

In the light of Lemma 4.3.3 we therefore see that the condition for $e^{-\tilde{H}t}$ to be bounded from L^2 to L^p is
$$1 - 2/p < (1-e^{-2t})/(1+e^{-2t})$$
which is equivalent to (4.3.3).

We next examine the spectrum of the generator $\tilde{H}_p$ of the semigroup $e^{-\tilde{H}t}$, regarded as acting on $L^p(\mathbb{R}, \phi_0^2 \, dx)$. Since $\tilde{H}_2$ is unitarily equivalent to H, we see that $\tilde{H}_2$ has compact resolvent and that
$$\text{Sp}(\tilde{H}_2) = \{0, 1, 2, 3, \ldots\}.$$
It follows that $\tilde{H}_p$ has compact resolvent with
$$\text{Sp}(\tilde{H}_p) = \{0, 1, 2, 3, \ldots\}$$
for all $1 < p < \infty$, by Theorem 1.6.3. However Theorem 1.6.4 is not applicable and we shall see that the spectrum of H_1 is quite different.

Theorem 4.3.5. *We have*
$$\text{Sp}(\tilde{H}_1) = \{z \in \mathbb{C} : \text{Re } z \geq 0\} \quad (4.3.4)$$

Indeed every z with $\operatorname{Re} z > 0$ is an eigenvalue of $\tilde{H}_1$ with multiplicity two.

Proof. We define the isometry V from $L^1(\mathbb{R}, dx)$ to $L^1(\mathbb{R}, \phi_0^2 dx)$ by
$$Vf = \phi_0^{-2} f$$
and then put
$$\hat{H} = V^{-1} \tilde{H}_1 V$$
so that the operator $\hat{H}$ on $L^1(\mathbb{R}, dx)$ has the same spectrum as $\tilde{H}_1$. The heat kernel $\hat{K}(t, x, y)$ of $e^{-\hat{H}t}$ is given by
$$\begin{aligned}\hat{K}(t, x, y) &= \phi_0(x)^2 \tilde{K}(t, x, y) \\ &= \phi_0(x) K(t, x, y) \phi_0(y)^{-1} \\ &= \{\pi(1 - e^{-2t})\}^{-\frac{1}{2}} \exp\{-B(t, x, y)\}\end{aligned} \quad (4.3.5)$$
where
$$B(t, x, y) = (x - e^{-t} y)^2 / (1 - e^{-2t}).$$

If $\mathscr{F}$ denotes the Fourier transform map from $L^1(\mathbb{R}, dx)$ to $C_0(\mathbb{R})$ then it follows that
$$(\mathscr{F} e^{-\hat{H}t} f)(k) = e^{-c_t k^2 / 4} f(e^{-t} k)$$
where
$$c_t = 1 - e^{-2t}.$$

Let $\chi^{\pm}$ be the characteristic functions of $\pm[0, \infty)$. Letting $f_z^{\pm}$ denote the L^1 functions whose Fourier transforms are $\chi^{\pm}(k) |k|^z e^{-k^2/4}$, we see that for any $\operatorname{Re} z > 0$
$$\begin{aligned}(\mathscr{F} e^{-Ht} f_z^{\pm})(k) &= \chi^{\pm}(k) e^{-c_t k^2 / 4} e^{-zt} |k|^z \exp(-e^{-2t} k^2 / 4) \\ &= e^{-zt} (\mathscr{F} f_z^{\pm})(k).\end{aligned}$$
Hence
$$e^{-Ht} f_z^{\pm} = e^{-zt} f_z^{\pm}$$
and
$$\hat{H} f_z^{\pm} = z f_z^{\pm}.$$

The validity of (4.3.4) now follows from the fact that $\operatorname{Sp}(\tilde{H}_1)$ is a closed subset of $\{z \in \mathbb{C} : \operatorname{Re} z \geq 0\}$, because $e^{-\tilde{H}_1 t}$ is a strongly continuous contraction semigroup.

Theorem 4.3.6. *If $1 < p < \infty$ then $\exp(-\tilde{H}_p t)$ is a norm analytic function of t for $0 < t < \infty$. However, for $p = 1$ we have*
$$\|e^{-\tilde{H}_1 s} - e^{-\tilde{H}_1 t}\| = 2 \quad (4.3.6)$$
whenever $0 < s < t < \infty$.

Proof. The first statement is a simple application of Theorem 1.4.2. We deduce (4.3.6) from the explicit form (4.3.5) of the heat kernel $\hat{K}$ of $e^{-\hat{H}t}$. If f

is any probability density in $L^1(\mathbb{R}, dx)$ and we put
$$f_a(x) = f(x - a)$$
then one sees that for any $0 < s < t < \infty$, $e^{-\hat{H}s} f_a$ and $e^{-\hat{H}t} f_a$ have asymptotically disjoint supports as $a \to \infty$ so
$$2 = \lim_{a \to \infty} \| e^{-\hat{H}s} f_a - e^{-\hat{H}t} f_a \|_1$$
$$\leqslant \| e^{-\hat{H}s} - e^{-\hat{H}t} \|.$$

4.4 Rosen's lemma

We now return to operators which satisfy all the assumptions of Theorem 4.2.1. Our goal will be to obtain conditions under which the semigroup $\exp(-H_\phi t)$ on $L^2(\Omega, \phi^2 \, dx)$ is ultracontractive.

Lemma 4.4.1. *Suppose that there exist constants a and c such that*
$$g \leqslant a \| g \|_{\mu/2} (H + c) \tag{4.4.1}$$
for some $2 < \mu < \infty$ and all $g \in L^{\mu/2}(\Omega)$. Suppose also that the quadratic form inequality
$$-\log \phi \leqslant \varepsilon H + \gamma(\varepsilon) \tag{4.4.2}$$
holds for some $\gamma(\varepsilon)$ and all $0 < \varepsilon < \infty$. Then
$$\log f \leqslant \varepsilon H + k - (\mu/4) \log \varepsilon + \varepsilon c/2 + \gamma(\varepsilon/2)$$
for all $0 \leqslant f \in L^2(\Omega, \phi^2 \, dx)$ of norm one, and all $0 < \varepsilon < \infty$.

Proof. If $\beta > 0$ we define the characteristic function χ by
$$\chi = \begin{cases} 1 & \text{if } \beta f \phi \geqslant 1, \\ 0 & \text{otherwise.} \end{cases}$$
Then
$$\log(\beta f \phi) \leqslant \chi \log(\beta f \phi)$$
$$\leqslant a \| \chi \log(\beta f \phi) \|_{\mu/2} (H + c).$$
Given μ there exists a constant $b > 0$ such that
$$(\log s)^{\mu/2} \leqslant b s^2$$
for all $s \geqslant 1$. This implies
$$\| \chi \log(\beta f \phi) \|_{\mu/2}^{\mu/2} \leqslant b \int (\chi \beta f \phi)^2 \leqslant b \beta^2.$$
Hence
$$\log(\beta f \phi) = a b^{2/\mu} \beta^{4/\mu} (H + c).$$

Putting
$$\varepsilon = 2ab^{2/\mu}\beta^{4/\mu}$$
we deduce that
$$\log f \leq -(\mu/4)\log \varepsilon + k_1 - \log \phi + (\varepsilon/2)(H+c)$$
The lemma follows by combining this with (4.4.2).

Corollary 4.4.2. *We have*
$$\int (f^2 \log f)\phi^2 \, dx \leq \varepsilon Q_\phi(f) + \beta(\varepsilon)\|f\|_2^2 + \|f\|_2^2 \log \|f\|_2 \quad (4.4.3)$$
for all $0 \leq f \in L^1 \cap L^\infty \cap \mathrm{Dom}(Q_\phi)$ *and all* $0 < \varepsilon < \infty$, *where*
$$\beta(\varepsilon) = k - (\mu/4)\log \varepsilon + \varepsilon c/2 + \gamma(\varepsilon/2).$$

Proof. By use of the unitary operator U_ϕ defined in Section 4.1 we obtain
$$\log f \leq \varepsilon H_\phi + \beta(\varepsilon)$$
for all $0 \leq f \in L^2(\Omega, \phi^2 \, dx)$ of norm one. The corollary follows immediately if $\|f\|_2 = 1$, and for general f by homogeneity arguments.

The conclusion (4.4.3) of this corollary is of precisely the right form for proving that $\exp(-H_\phi t)$ is ultracontractive; see Section 2.2 and particularly Corollary 2.2.8. Of course, it is crucial for applications that $\beta(\varepsilon)$ should not diverge too rapidly as $\varepsilon \to 0$.

The hypothesis (4.4.1) is of a standard type which we have already analysed in Theorem 2.4.2 and Theorem 2.4.5. Note that this condition refers to H and not to H_ϕ. The requirement that $\mu > 2$ is not essential and can be circumvented either by reorganising the proof slightly or by using the idea in the proof of Theorem 2.4.4.

The quadratic form inequality (4.4.2) may be investigated using the techniques of Section 1.5. In the case of Schrödinger operators we can also use subharmonic comparison inequalities, and we now describe this approach in more detail.

We assume that $H = -\Delta + V$ on $L^2(\Omega)$ where V lies in $L^1_{\mathrm{loc}}(\Omega)$ and that ϕ is a positive C^2 function on Ω such that
$$-\Delta\phi + V\phi \geq 0.$$

Lemma 4.4.3. *Let* $W:\Omega \to \mathbb{R}$ *be* C^2 *and satisfy:*

(i) $$W \leq \varepsilon H + \gamma(\varepsilon)$$
for some $\gamma(\varepsilon)$ *and all* $0 < \varepsilon < \infty$:

(ii) $$W(x) \to +\infty \quad \text{as} \quad x \to \partial\Omega \cup \{\infty\}:$$

(iii) $$|\nabla W|^2 - \Delta W \geq V_+$$

outside some compact subset K of Ω;

(iv) $\qquad e^{-W} \leqslant \phi$

inside the set K.
Then
$$-\log \phi \leqslant \varepsilon H + \gamma(\varepsilon)$$
for all $0 < \varepsilon < \infty$.

Proof. We put $U = \Omega \backslash K$ and $\psi = e^{-W}$ so that $\psi \leqslant \phi$ on $\partial U \cup \{\infty\}$. Also
$$-\Delta \psi + X\psi = 0$$
in U, where
$$X = |\nabla W|^2 - \Delta W \geqslant V_+.$$
The following subharmonic comparison theorem implies that $\psi \leqslant \phi$ throughout U, and therefore in Ω. Therefore
$$-\log \phi \leqslant W \leqslant \varepsilon H + \gamma(\varepsilon)$$
as required.

Proposition 4.4.4. *Let U be a region in $\mathbb{R}^N$ and let ϕ, ψ be two positive C^2 functions on U such that*

(i) $\qquad \psi \leqslant \phi$ on $\partial U \cap \{\infty\}$;

(ii) $\qquad -\Delta \phi + V\phi \geqslant 0 \quad$ in U;

(iii) $\qquad -\Delta \psi + W\psi \leqslant 0 \quad$ in U;

(iv) $\qquad V_+ \leqslant W \qquad\qquad$ in U.

Then $\psi \leqslant \phi$ throughout U.

Proof. If $h = \psi - \phi$ and x lies in $D = \{x : h(x) > 0\}$ then
$$\begin{aligned}\Delta h &= \Delta \psi - \Delta \phi \\ &\geqslant W\phi - V\phi \\ &\geqslant W\psi - W\phi \\ &= Wh > 0.\end{aligned}$$
Since h is subharmonic in D with $h \leqslant 0$ on $\partial D \cup \{\infty\}$, we conclude that $h \leqslant 0$ on D, which must therefore be empty.

4.5 Schrödinger operators

In this section we apply the above methods to the Schrödinger operator $H = -\Delta + V$ on $L^2(\mathbb{R}^N)$, where we assume that the potential V is continuous and that $V(x) \to +\infty$ as $|x| \to \infty$; the analysis when V has local singularities is much harder and is deferred to Section 4.8. Under the above conditions the resolvent is compact, and H has a strictly positive C^2 ground state ϕ which we normalise by $\|\phi\|_2 = 1$. We let E denote the corresponding eigenvalue.

Lemma 4.5.1. *Suppose that there exist constants $a_i > 2$ and $c_i > 0$ such that*

$$c_1|x|^{a_1} - c_2 \leqslant V(x) - E \leqslant c_3|x|^{a_2} + c_4 \qquad (4.5.1)$$

for all $x \in \mathbb{R}^N$ where $a_2 < 2a_1 - 2$, and that W is defined by

$$W(x) = |x|^a$$

where

$$a_2 < 2a - 2 < 2a_1 - 2. \qquad (4.5.2)$$

Then

$$W \leqslant \varepsilon(H - E) + c_2\varepsilon + c_5\varepsilon^{-a/(a_1-a)}$$

for all $\varepsilon > 0$.

Proof. A straightforward computation shows that

$$r^a \leqslant \varepsilon r^{a_1} + c_6\varepsilon^{-a(a_1-a)}$$

for all $\varepsilon > 0$ and $r \geqslant 0$. Therefore

$$\begin{aligned}
W &\leqslant \varepsilon(c_1|x|^{a_1} - c_2) + c_2\varepsilon + c_6(c_1\varepsilon)^{-a/(a_1-a)} \\
&\leqslant \varepsilon(V - E) + c_2\varepsilon + c_5\varepsilon^{-a/(a_1-a)} \\
&\leqslant \varepsilon(H - E) + c_2\varepsilon + c_5\varepsilon^{-a/(a_1-a)}.
\end{aligned}$$

Lemma 4.5.2. *We have*

$$-\log \phi \leqslant \varepsilon(H - E) + \gamma(\varepsilon)$$

for all $0 < \varepsilon < \infty$, where

$$\gamma(\varepsilon) = c + c_2\varepsilon + c_5\varepsilon^{-a/(a_1-a)}.$$

Proof. A straightforward computation shows that

$$|\nabla W|^2 - \Delta W = a^2|x|^{2a-2} - a(a + N - 2)|x|^{a-2}$$

for all $x \in \mathbb{R}^N$. It follows that there exists $R > 0$ such that $|x| > R$ implies

$$|\nabla W|^2 - \Delta W \geqslant (V - E)_+.$$

Moreover if $c > 0$ is large enough then the Harnack inequality implies that $e^{-W} \leqslant e^c\phi$ whenever $|x| \leqslant R$. Lemma 4.4.3 now implies that

$$-\log(e^c\phi) \leqslant \varepsilon(H - E) + c_2\varepsilon + c_5\varepsilon^{-a/(a_1-a)}$$

which yields the lemma immediately.

Proposition 4.5.3. *If $N > 2$ then the operator*
$$\tilde{H} = U_\phi^{-1}(H - E)U_\phi$$
on $L^2(\mathbb{R}^N, \phi^2 \, dx)$ satisfies the logarithmic Sobolev inequality
$$\int (f^2 \log f)\phi^2 \, dx \leq \varepsilon \tilde{Q}(f) + \beta(\varepsilon)\|f\|_2^2 + \|f\|_2^2 \log \|f\|_2$$
for all $0 < \varepsilon < \infty$ and $0 \leq f \in L^1 \cap L^\infty \cap \text{Quad}(\tilde{H})$, where
$$\beta(\varepsilon) = k_1 - (N/4)\log \varepsilon + \varepsilon(c_2 + E/2) + c_6 \varepsilon^{-a/(a_1 - a)}.$$

Proof. Since V is bounded below on $\mathbb{R}^N$ by $-c_2$ the method of Example 2.1.9 leads to
$$\|e^{-(H+c_2)t}\|_{\infty,1} \leq (4\pi t)^{-N/2}$$
for all $t > 0$. Theorem 2.4.5 now implies that
$$g \leq a\|g\|_{N/2}(H + c_2)$$
for all $g \in L^{N/2}(\mathbb{R}^N)$. The proof is completed by applying Corollary 4.4.2.

Theorem 4.5.4. *If V satisfies the hypothesis (4.5.1) of Lemma 4.5.1 and a satisfies (4.5.2) then*
$$0 \leq K(t, x, y) \leq b_1 \exp(b_2 t^{-b})\phi(x)\phi(y)$$
for all $x, y \in \mathbb{R}^N$ and $0 < t \leq 1$ where $b_i > 0$ and
$$1 < b = a/(a_1 - a) < \infty. \tag{4.5.3}$$

Proof. We first consider the case $N > 2$. Proposition 4.5.3 implies that
$$\int (f^2 \log f)\phi^2 \, dx \leq \varepsilon \tilde{Q}(f) + \beta_1(\varepsilon)\|f\|_2^2 + \|f\|_2^2 \|f\|_2$$
for all $0 < \varepsilon \leq 1$, where
$$\beta_1(\varepsilon) = b_3 \varepsilon^{-b}.$$
Note that (4.5.1) implies $a_1 \leq a_2$ and in combination with (4.5.2) this implies (4.5.3). The application of Example 2.3.4 now yields
$$0 \leq \tilde{K}(t, x, y) \leq b_4 \exp(b_2 t^{-b})$$
for $0 < t \leq 1$. The proof is completed by using the formula
$$\tilde{K}(t, x, y) = e^{Et}\phi(x)^{-1}K(t, x, y)\phi(y)^{-1}$$
which follows as in Lemma 4.2.2.

If $N \leq 2$ then we consider instead the operator
$$H_0 = H \otimes 1 \otimes 1 + 1 \otimes H \otimes 1 + 1 \otimes 1 \otimes H$$

on $L^2(\mathbb{R}^{3N})$. Since the ground state and heat kernel of H_0 decompose as direct products, the same conclusion holds.

Corollary 4.5.5. *Let* $H = -\Delta + |x|^\alpha$ *on* $L^2(\mathbb{R}^N)$ *where* $2 < \alpha < \infty$ *and let*

$$(\alpha + 2)/(\alpha - 2) < b < \infty. \qquad (4.5.4)$$

Then the heat kernel of e^{-Ht} *satisfies*

$$0 \leq K(t, x, y) \leq b_1 \exp(b_2 t^{-b})\phi(x)\phi(y)$$

for all $x, y \in \mathbb{R}^N$ *and* $0 < t \leq 1$, *where* ϕ *is the ground state of* H.

Proof. The condition (4.5.2) with $a_1 = a_2 = \alpha$ is equivalent to (4.5.4), and the corollary is now a special case of Theorem 4.5.4.

The above corollary is sharp in the sense that if $\alpha = 2$ then no such bound on the heat kernel exists, because of Theorem 4.3.2. We can, however, also show that the restriction (4.5.4) on the possible values of b is sharp. This requires some preparation. We start by using the subharmonic comparison inequality (Proposition 4.4.4) to obtain sharp bounds on the ground state ϕ of H, whose corresponding eigenvalue is denoted by E.

Lemma 4.5.6. *Let* $\lambda \in \mathbb{R}$ *and put*

$$\beta = \alpha/4 + (N - 1)/2.$$

Then the radial function

$$f_\lambda(r) = r^{-\beta} \exp\left\{-\frac{2}{2+\alpha} r^{1+\alpha/2} - \frac{\lambda}{\alpha - 2} r^{1-\alpha/2}\right\}$$

satisfies

$$\Delta f_\lambda / f_\lambda = r^\alpha - \lambda + \frac{\beta(\beta + 2 - N)}{r^2} + \frac{\lambda^2}{4r^\alpha} - \frac{\lambda\alpha}{2r^{\alpha/2+1}}.$$

Proof. This is a direct computation. Putting

$$f(r) = r^{-\beta} e^{-g(r)}$$

we have

$$\Delta f = f''(r) + \frac{N-1}{r} f'(r)$$

so

$$\Delta f / f = \beta(\beta + 1)r^{-2} + 2\beta g' r^{-1} + (g')^2 - g'' - (N-1)\beta r^{-2} - (N-1)g' r^{-1}.$$

Substituting

$$g'(r) = r^{\alpha/r} - (\lambda/2)r^{-\alpha/2}.$$

into this yields the result.

Lemma 4.5.7. *If $\alpha > 0$ then there exist constants $c_i > 0$ such that*
$$c_1 f_0(|x|) \leqslant \phi(x) \leqslant c_2 f_{2E}(|x|)$$
for all large enough $|x|$.

Proof. We have
$$-\Delta\phi + (V - E)\phi = 0$$
and
$$-\Delta f_{2E} + W f_{2E} = 0$$
where
$$W = V - 2E + o(1)$$
as $|x| \to \infty$. Therefore $V - E \geqslant W \geqslant 0$ for large enough $|x|$, say $|x| \geqslant R$. If $|x| = R$ then $\phi \leqslant c_2 f_{2E}$ for some $c_2 < \infty$ by compactness. The second inequality of the lemma is now a direct consequence of Proposition 4.4.4. The other inequality has a similar proof.

It is remarkable that the above bound becomes sharp if $\alpha > 2$, and this is related to the ultracontractivity of $e^{-\tilde{H}t}$.

Corollary 4.5.8. *If $\alpha > 2$ then there exist constants $c_i > 0$ such that*
$$c_1 f_0(|x|) \leqslant \phi(x) \leqslant c_2 f_0(|x|)$$
for all large enough $|x|$.

Proof. We simply observe that f_0 and f_{2E} are comparable as $r \to \infty$.

We next obtain a lower bound on the heat kernel K of the operator $H = -\Delta + |x|^\alpha$.

Lemma 4.5.9. *If $\alpha > 0$ then there exist positive constants c and T_5 such that*
$$K(t, x, x) \geqslant c t^{-N/2} \exp\{-(|x| + 1)^\alpha t\}$$
for all $0 < t \leqslant T_5$ and $x \in \mathbb{R}^N$.

Proof. If H_B is the operator obtained from H by imposing Dirichlet boundary conditions on the surface of the ball B with centre x and radius 1, then
$$K(t, x, x) \geqslant K_B(t, x, x)$$
for all $t > 0$. Moreover
$$|y|^\alpha \leqslant (|x| + 1)^\alpha$$
for all $y \in B$ so
$$K_B(t, x, x) \geqslant K_0(t, x, x) \exp\{-(|x| + 1)^\alpha t\}$$

for all $t>0$, where K_0 is the heat kernel of $-\Delta$ on B with Dirichlet boundary conditions. The proof is completed by applying Lemma 3.3.3 with $T_5 \beta^2 = t$ and $0 < \beta \leqslant 1$.

Theorem 4.5.10. *If $\alpha > 2$ and the heat kernel of $H = -\Delta + |x|^\alpha$ satisfies*
$$K(t,x,y) \leqslant c(t)\phi(x)\phi(y)$$
for all $t>0$ and $x, y \in \mathbb{R}^N$, then the exist $T>0$ and $c_i > 0$ such that
$$c(t) \geqslant c_1 \exp\{c_2 t^{-(\alpha+2)/(\alpha-2)}\}$$
for all $0 < t \leqslant T$.

Proof. By combining Corollary 4.5.8 and Lemma 4.5.9 we see that $0 < t \leqslant T_5$ implies
$$ct^{-N/2}\exp\{-(r+1)^\alpha t\} \leqslant c(t)r^{-2\beta}\exp\left\{-\frac{4}{2+\alpha}r^{1+\alpha/2}\right\}$$
provided r is large enough, say $r \geqslant R \geqslant 1$. We now put $T = T_5 \wedge T_6$ where T_6 is defined by
$$R = (2^{\alpha-1}\alpha T_6)^{-2/(\alpha-2)}.$$
Then $0 < t \leqslant T$ and
$$r = (2^{\alpha-1}\alpha t)^{-2/(\alpha-2)}$$
implies $r \geqslant R$. Hence
$$c(t) \geqslant ct^{-N/2}r^{2\beta}\exp\left\{-(r+1)^\alpha t + \frac{4}{2+\alpha}r^{1+\alpha/2}\right\}$$
$$\geqslant ct^{-N/2}\exp\left\{-2^\alpha r^\alpha t + \frac{4}{2+\alpha}r^{1+\alpha/2}\right\}$$
$$= ct^{-N/2}\exp\{c_3 t^{-(\alpha+2)/(\alpha-2)}\} \tag{4.5.5}$$
where
$$c_3 = \frac{4}{2+\alpha}(2^{\alpha-1}\alpha)^{-(\alpha+2)/(\alpha-2)} - 2^\alpha(2^{\alpha-1}\alpha)^{-2\alpha/(\alpha-2)}$$
$$= \left(\frac{4}{2+\alpha} - \frac{2}{\alpha}\right)(2^{\alpha-1}\alpha)^{-(\alpha+2)/(\alpha-2)}$$
$$= 2\frac{\alpha-2}{\alpha+2}(2^{\alpha-1}\alpha)^{-(\alpha+2)/(\alpha-2)}$$
$$> 0.$$

Elliptic operators on bounded regions

The factor $t^{-N/2}$ in (4.5.5) may be absorbed into the exponential by decreasing the constant c_3 slightly.

Theorem 4.5.10 proves that the lower bound on b given in (4.5.4) is sharp, and once again illustrates the power of logarithmic Sobolev inequalities. The rate at which $c(t)$ diverges as $t \to 0$ shows that there is no hope of proving ultracontractivity of $e^{-\tilde{H}t}$ by the use of ordinary Sobolev or Nash inequalities as described in Section 2.4.

We finish the section by stating a theorem which shows that if $H = -\Delta + |x|^\alpha$ for $0 < \alpha < 2$ then e^{-Ht} is not even hypercontractive.

Theorem 4.5.11. *Let $H = -\Delta + V$ on $L^2(\mathbb{R}^N)$ where $V \in L^1_{\text{loc}}$ is bounded below. Suppose that $H\phi = E\phi$ where $0 < \phi \in L^2(\mathbb{R}^N)$ and put*

$$\tilde{H} = U_\phi^{-1}(H - E)U_\phi$$

where $U_\phi : L^2(\mathbb{R}^N, \phi^2 \, dx) \to L^2(\mathbb{R}^N, dx)$ is given by $U_\phi f = \phi f$. Then if $e^{-\tilde{H}t}$ is hypercontractive there exist constants $\gamma > 0$ and $\beta \in \mathbb{R}$ such that

$$H \geqslant \gamma x^2 - \beta$$

as a quadratic form inequality on $L^2(\mathbb{R}^N)$.

Clearly the conclusion of this theorem is not valid for $V(x) = |x|^\alpha$ if $0 \leqslant \alpha < 2$. For a related result which gives a condition under which $e^{-\tilde{H}t}$ is not ultracontractive see Corollary 4.7.4.

4.6 Elliptic operators on bounded regions

In this section we suppose that Ω is a bounded open region in $\mathbb{R}^N$ and that H is defined on $C_c^2(\Omega)$ by

$$Hf = -\sum \frac{\partial}{\partial x_i}\left(a_{ij} \frac{\partial f}{\partial x_j}\right)$$

where a is a positive C^1 matrix-valued function on $\bar{\Omega}$, so that H is uniformly elliptic. We extend H to a self-adjoint operator on $L^2(\Omega)$ satisfying Dirichlet boundary conditions by Theorem 1.2.5. Then H^{-1} is compact by Theorem 1.6.8. The lowest eigenvalue E of H has multiplicity one by the Harnack inequality (Corollary 3.3.7) and the corresponding eigenfunction ϕ is strictly positive on Ω, and C^2 by local elliptic regularity theorems. Moreover, ϕ vanishes weakly on $\partial\Omega$ in the sense that $\phi \in W_0^{1,2}(\Omega)$; see Theorems 1.5.6 and 1.5.7 for further discussion of this.

We say that $\partial\Omega \in C^k$ if there exists a function $\rho \in C^k(\mathbb{R}^N)$ such that

$$\Omega = \{x : \rho(x) > 0\}$$

and there exists $v > 0$ such that

$$v^{-1} \leq |\nabla \rho| \leq v \tag{4.6.1}$$

for all x in some neighbourhood of $\partial \Omega$. If $k \geq 1$ and such a ρ exists then there exists $c > 0$ such that

$$c^{-1} d(x) \leq \rho(x) \leq c d(x) \tag{4.6.2}$$

for all $x \in \Omega$, by an easy local argument in a neighbourhood of $\partial \Omega$.

We start with a version of the Hopf boundary point lemma.

Lemma 4.6.1. *If $\partial \Omega$ is C^2 then there exists $a_0 > 0$ such that the ground state ϕ of H satisfies $\phi \geq a_0 d$.*

Proof. We have $\rho^2 \in \mathrm{Dom}(H)$ with

$$H(\rho^2)(x) = -2\rho(x)\alpha(x) - 2 \sum a_{ij}(x) \frac{\partial \rho}{\partial x_i} \frac{\partial \rho}{\partial x_j} \tag{4.6.3}$$

where

$$\alpha(x) = \sum \left(\frac{\partial a_{ij}}{\partial x_i} \frac{\partial \rho}{\partial x_j} + a_{ij} \frac{\partial^2 \rho}{\partial x_i \partial x_j} \right).$$

Since α is bounded the first term on the RHS of (4.6.3) converges to zero as $x \to \partial \Omega$. By combining (4.6.1) and the uniform ellipticity there exist $\varepsilon > 0$ and $\lambda > 0$ such that

$$H(\rho^2)(x) \leq -\lambda$$

whenever $d(x) < \varepsilon$. Hence there exists $\mu > 0$ such that

$$H(\rho + \mu^2 \rho^2)(x) \leq -1$$

whenever $d(x) < \varepsilon$. The Harnack inequality now implies that

$$\phi \geq a_1 H(\rho + \mu^2 \rho^2)$$

on Ω for some $a_1 > 0$. Since H^{-1} is positivity-preserving we deduce using (4.6.2) that

$$E^{-1} \phi = H^{-1} \phi \geq a_1 (\rho + \mu^2 \rho^2) \geq a_1 c^{-1} d.$$

Theorem 4.6.2. *If $\partial \Omega$ is C^2 then the heat kernel K of e^{-Ht} satisfies*

$$0 \leq K(t, x, y) \leq c t^{-1(1 + N/2)} \phi(x) \phi(y) \tag{4.6.4}$$

for all $x, y \in \Omega$ and all $t > 0$.

Proof. We verify the conditions of Corollary 4.4.2. The Condition (4.4.1) holds with $\mu = N$ and $c = 0$ by Theorems 2.3.6 and 2.4.5. Lemma 4.6.1 and

Theorem 1.5.5 together imply that
$$-\log \phi \leqslant -\log a_0 - \log d$$
$$\leqslant -\log a_0 + \varepsilon d^{-2} - \tfrac{1}{2}\log \varepsilon$$
$$\leqslant -\log a_0 + \varepsilon a_1 H - \tfrac{1}{2}\log \varepsilon$$
for all $\varepsilon > 0$. Corollary 4.4.2 now holds with
$$\beta(\varepsilon) = a_2 - (N/4 + \tfrac{1}{2})\log \varepsilon$$
and the theorem follows by applying Example 2.3.1.

We comment without elaboration that the half-line example in Section 4.1 indicates that the power of t in (4.6.4) is sharp.

Corollary 4.6.3. *If $\partial\Omega$ is C^2 and $H\phi_n = E_n\phi_n$ where $\|\phi_n\|_2 = 1$ then*
$$|\phi_n(x)| \leqslant cE_n^{1/2+N/4}\phi(x)$$
for all $x \in \Omega$.

Proof. We can apply Theorem 4.2.4 with
$$c_t = c(t^{-(1+N/2)} \vee 1).$$
We obtain (4.2.3) with
$$c_n = \inf\{c_t \exp(E_n - E)t : 0 < t < \infty\}$$
and putting $t = E_n^{-1}$ yields the result.

If $\partial\Omega$ is not C^2 then Lemma 4.6.1 is typically false, and the behaviour of the ground state ϕ near $\partial\Omega$ can be quite complicated.

Example 4.6.4. If $\Omega = (0, 1) \times (0, 1) \subseteq \mathbb{R}^2$ then
$$\phi(x, y) = c \sin(\pi x)\sin(\pi y)$$
and $\phi(x, y) \sim d(x, y)$ as (x, y) converges to any point of $\partial\Omega$ other than a vertex. However, if $y = x^\gamma$ where $\gamma \geqslant 1$ then
$$\phi(x, y) \sim x^{1+\gamma} \sim d(x, y)^{(1+\gamma)/\gamma}$$
as $x \to 0$, and the best uniform bounds on ϕ in terms of d alone are of the form
$$c_1 d^2 \leqslant \phi \leqslant c_2 d$$
where $c_i > 0$.

Our next example clarifies the nature of the boundary behaviour for a piecewise smooth polygonal region in $\mathbb{R}^2$.

Example 4.6.5. If $0 < \alpha \leqslant 2\pi$ we define $\Omega_\alpha \subseteq \mathbb{R}^2$ by
$$\Omega_\alpha = \{re^{i\theta} : 0 < r < 1 \text{ and } 0 < \theta < \alpha\}.$$

The ground state is of the form
$$\phi(re^{i\theta}) = cR(r)\sin(\pi\theta/\alpha)$$
where $R(r) > 0$ for $0 < r < 1$, $R(0) = R(1) = 0$ and
$$-R'' - \frac{1}{r}R' + \frac{\pi^2}{\alpha^2 r^2}R = E_\alpha R$$

E_α being the smallest eigenvalue of $H = -\Delta$ in Ω_α. Standard calculations establish that
$$R(r) \sim r^{\pi/\alpha}$$
as $r \to 0$. Hence
$$\phi(z) \sim |z|^{\pi/\alpha}$$
as $z \to 0$ non-tangentially.

Example 4.6.6. Let $\Omega \subseteq \mathbb{R}^2$ be defined by
$$\Omega = \{(x, y) : 0 < x < 1, |y| < x\}$$
and for $0 < \lambda < \infty$ let
$$H = -\frac{\partial^2}{\partial x^2} - \frac{1}{\lambda^2}\frac{\partial^2}{\partial y^2}$$
subject to Dirichlet boundary conditions on $\partial\Omega$. Putting $y = u/\lambda$ we see that H_λ is unitarily equivalent to the operator $-\Delta$ on the region
$$\Omega_\lambda = \{(u, x) : 0 < x < 1 \text{ and } |u| < x\lambda\}.$$
Now the ground state ψ_λ of this operator satisfies
$$\psi_\lambda(z) \sim |z|^\alpha$$
as $z \to 0$ non-tangentially where
$$\alpha = 2\arctan\lambda$$
by a comparison argument with Example 4.6.5. Therefore the ground state ϕ_λ of H_λ on Ω satisfies
$$\phi_\lambda(z) \sim |z|^\alpha$$
as $z \to 0$ non-tangentially. The point of this example is to show that although the various operators H_λ are all uniformly elliptic with constant coefficients their ground states vanish at different rates as one approaches a vertex on the boundary. This is in strong contrast to the situation discussed earlier where one has a C^2 boundary.

The next example will be important in our subsequent analysis.

Example 4.6.7. We define the conical region $U \subseteq \mathbb{R}^N$ in polar coordinates

by
$$U = \{(r,\omega): 0 < r < \delta \quad \text{and} \quad \omega \in A\}$$
where A is an open subset of the unit sphere S^{N-1} with smooth boundary. Given $\lambda \geq 0$ we define the function f_U on U by
$$f_U(r,\omega) = R(r)S_A(\omega)$$
where $S_A > 0$ is the ground state of the Laplace–Beltrami operator Δ_A on A with Dirichlet boundary conditions and $-\Delta_A S_A = E_A S_A$; also R is the solution of
$$-R'' - (N-1)r^{-1}R' + E_A r^{-2} R = \lambda R \tag{4.6.5}$$
on $(0,\delta)$ such that $R(0) = 0$. A direct calculation shows that
$$-\Delta f_U = \lambda f_U$$
on U, and that $f_U = 0$ on $\partial U \cap \{(r,\omega): r < \delta\}$. If ω is restricted to a compact subset of A, then an asymptotic analysis of (4.6.5) shows that
$$f_U(r,\omega) \sim r^\alpha$$
as $r \to 0$, where α is the positive solution of
$$\alpha(\alpha + N - 2) = E_A. \tag{4.6.6}$$
If λ is chosen to be the smallest value for which one has $R(\delta) = 0$, then f_U is the ground state of U, and the above calculations determine the asymptotic behaviour of the ground state near the vertex of the cone U. If, however, $\lambda \geq 0$ is given then for small enough δ we will have $R(\delta) > 0$, and hence $f_U > 0$ on U and on $\partial U \cap \{(r,\omega): r = \delta\}$; if $\lambda = 0$ the smallness of δ need not be assumed.

We now turn to the study of the heat kernel and ground state of a uniformly elliptic operator in a bounded region Ω for which $\partial \Omega$ is not C^2. We have indicated that their behaviour depends in a very complicated way upon both $\partial \Omega$ and the coefficients $a_{ij}(x)$. To simplify matters we shall therefore only consider the case where
$$a_{ij}(x) = \delta_{ij}$$
so that $H = -\Delta$ on $L^2(\Omega)$ with Dirichlet boundary conditions.

If $x, y \in \mathbb{R}^N$ and $\theta > 0$, $\delta > 0$, we define a conical region by
$$C(y, x, \theta, \delta) = \left\{ z \in \mathbb{R}^N : |z - y| < \delta \quad \text{and} \quad \frac{(z-y)\cdot(x-y)}{|z-y||x-y|} > \cos\theta \right\}.$$
We define its base by
$$\partial_1 C = \{z \in \partial C : |z - y| = \delta\}.$$
We say that a bounded region Ω satisfies an internal cone condition if there exist $\theta > 0$ and $\delta > 0$ such that for all $x \in \Omega$ there exists $y_x \in \partial \Omega$

with

$$C(y_x, x, \theta, \delta) \subseteq \Omega.$$

By reducing θ and δ slightly we may also assume that

$$\partial_1 C(y_x, x, \theta, \delta) \subseteq B$$

where B is a fixed compact subset of Ω. The external cone condition is defined similarly.

Theorem 4.6.8. *Let Ω be bounded and satisfy an internal cone condition. Then the ground state ϕ of $H = -\Delta$ on $L^2(\Omega)$ subject to Dirichlet boundary conditions satisfies*

$$\phi \geq a d^\alpha$$

for some $a > 0$ and $\alpha > 0$.

Proof. Let θ, δ, B be as in the definition above. Then the Harnack inequality implies that there exists $\mu > 0$ such that $\phi \geq \mu$ on B. Let $x \notin B$, put $U = C(y_x, x, \theta, \delta)$ and let f_U be defined as in Example 4.6.7, subject to the normalizations $\|f_U\|_\infty = 1$ and $\lambda = 0$. Then $-\Delta \phi = E \phi$ and $-\Delta f_U = 0$ on U; also $\phi \geq \mu f_U$ on $\partial_1 U$ and hence on the whole of ∂U. From $\Delta(\phi - \mu f_U) \leq 0$ in U and $\phi - \mu f_U \geq 0$ on ∂U we deduce that $\phi - \mu f_U \geq 0$ in U. Applying the results of Example 4.6.7 with α given by (4.6.6) we obtain

$$\phi(x) \geq \mu f_U(x) \geq \mu c_1 |x - y_x|^\alpha \geq c d(x)^\alpha.$$

A similar bound for $x \in B$ follows straight from the Harnack inequality.

Theorem 4.6.9. *Let Ω be bounded and satisfy internal and external cone conditions. Then the heat kernel K of $e^{\Delta t}$ subject to Dirichlet boundary conditions satisfies*

$$0 \leq K(t, x, y) \leq c_\delta t^{-\mu/2} \phi(x) \phi(y) \exp\{-(x-y)^2/4(1+\delta)t\}$$

for all $x, y \in \Omega$ and all $t > 0$, where $\mu \geq N$, ϕ is the ground state of $-\Delta$, and $0 < \delta < 1$ is arbitrary.

Proof. We verify the conditions of Corollary 4.4.2. The condition (4.4.1) holds with $\mu = N$ and $c = 0$ by Theorems 2.3.6 and 2.4.5. Theorems 4.6.8 and 1.5.5 together imply that

$$-\log \phi \leq -\log a - \log d^\alpha$$
$$\leq -\log a + \alpha \varepsilon d^{-2} - (\alpha/2) \log \varepsilon$$
$$\leq -\log a + \varepsilon a_1 H - (\alpha/2) \log \varepsilon.$$

Therefore Corollary 4.4.2 holds with

$$\beta(\varepsilon) = a_2 - (N/4 + \alpha/2)\log\varepsilon.$$

The ultracontractivity of $\exp(-H_\phi t)$ follows by applying Example 2.3.1 with N replaced by $\mu = N + 2\alpha$. The proof is completed by an adaptation of the argument leading to Corollary 3.2.8. The replacement of the Lebesgue measure dx by $\phi^2\,dx$ throughout that section causes no new problems.

Note 4.6.10. Both Theorem 4.6.8 and Theorem 4.6.9 remain valid if $-\Delta$ is replaced by a general uniformly elliptic operator in divergence form. However, because the proof depends upon the Harnack inequality, it is difficult to obtain useful bounds upon the constants α or μ.

Using the above theorems we are now able to prove a partial extension of Example 4.1.3 to more general regions.

Theorem 4.6.11. *Let $N \geqslant 3$ and let Ω be a bounded region in $\mathbb{R}^N$ which satisfies internal and external cone conditions. If ϕ is the ground state of $H = -\Delta$ subject to Dirichlet boundary conditions, then the Green function satisfies*

$$a_N^{-1}\phi(x)\phi(y) \leqslant G(x,y) \leqslant a_N\left(\frac{1}{|x-y|^{N-2}} \wedge \frac{\phi(x)\phi(y)}{|x-y|^{\mu-2}}\right)$$

for some $a_N > 0$ and $\mu \geqslant N$.

Proof. The lower bound on G follows from Theorem 4.2.5. The upper bound

$$G(x,y) \leqslant a_N |x-y|^{2-N}$$

follows from Corollary 3.2.8 and Lemma 3.4.3. Putting

$$\tilde{G}(x,y) = \phi(x)^{-1}\phi(y)^{-1}G(x,y)$$

the upper bound

$$\tilde{G}(x,y) \leqslant a_N |x-y|^{2-\mu}$$

follows from Theorem 4.6.9 and Lemma 3.4.3, this time acting on $L^2(\Omega, \phi^2\,dx)$.

4.7 Singular elliptic operators

Let a and b be two positive functions on the region $\Omega \subseteq \mathbb{R}^N$ such that $a^{\pm 1}$ and $b^{\pm 1}$ are all locally bounded. Then we may define a quadratic form on $C_c^\infty(\Omega)$ by

$$Q(f) = \int_\Omega a|\nabla f|^2. \qquad (4.7.1)$$

We regard C_c^∞ as a subspace of $L^\infty(\Omega, b\,dx)$, so that

$$\|f\|^2 = \int_\Omega b|f|^2. \tag{4.7.2}$$

By the argument of Theorem 1.2.6 the form Q is closable and we define $H \geq 0$ to be the self-adjoint operator associated with its closure. If $a \in W^{1,2}_{\text{loc}}(\Omega)$ then

$$Hf = -b^{-1}\nabla\cdot(a\nabla f)$$
$$= -b^{-1}a\Delta f - b^{-1}\nabla a\cdot\nabla f$$

for all $f \in C_c^\infty \subseteq \text{Dom}(H)$, but we do not make this hypothesis below.

The study of how the spectral properties of H depend upon a and b is a fascinating and complicated topic. If $b = 1$ then three simple results are given in Theorems 1.5.14, 1.6.6 and Corollary 1.6.7. We start this section by considering the case where

$$a = b = \phi^2.$$

Although this is merely a reformulation of the results of Section 4.5, it concentrates on the function ϕ instead of the potential V. As well as being technically simpler to prove the results, this is more natural if one wishes to think in probabilistic rather than quantum-mechanical terms.

We start with a positive C^2 function ϕ on $\mathbb{R}^N$ and define the operator H_ϕ on $C_c^1(\mathbb{R}^N)$ by the formal expression

$$H_\phi f = -\Delta f - 2\nabla(\log \phi)\cdot\nabla f.$$

If we regard C_c^1 as a dense subspace of the weighted Hilbert space $L^2(\mathbb{R}^N, \phi^2\,dx)$ then the associated quadratic form is given by

$$Q_\phi(f) = \int |\nabla f|^2 \phi^2 \, dx$$

which is non-negative. We extend H_ϕ to be a non-negative self-adjoint operator by taking its Friedrichs extension (Theorem 1.2.8). It follows by Lemma 1.3.4 that $\exp(-H_\phi t)$ is a symmetric Markov semigroup on $L^2(\mathbb{R}^N, \phi^2\,dx)$.

Theorem 4.7.1. *Suppose that $\phi = e^{-\psi}$ where $\psi \in C^2$ is bounded below, and that there exist $c_i > 0$ and $\delta > 0$ such that*

$$\psi \leq \varepsilon(|\nabla\psi|^2 - \Delta\psi) + c_1 + c_2\varepsilon^{-\delta} \tag{4.7.3}$$

for all $0 < \varepsilon < \infty$. Then there exist $a_i > 0$ and $\gamma > 0$ such that the heat kernel K_ϕ satisfies

$$0 \leq K_\phi(t, x, y) \leq a_1 \exp(a_2 t^{-\gamma})$$

for all $0 < t \leq 1$ and $x, y \in \mathbb{R}^N$.

Singular elliptic operators

Proof. We put
$$V = \Delta\phi/\phi = |\nabla\psi|^2 - \Delta\psi$$
so that
$$(-\Delta + V)\phi = 0.$$
Putting $\varepsilon = 1$ in (4.7.3) we see that V is continuous and bounded below. Moreover H_ϕ is unitarily equivalent to $H = -\Delta + V$ and (4.7.3) can be rewritten in the form
$$-\log\phi \leqslant \varepsilon V + c_1 + c_2\varepsilon^{-\delta}$$
$$\leqslant \varepsilon H + c_1 + c_2\varepsilon^{-\delta}.$$
The remainder of the proof follows that of Proposition 4.5.3 and Theorem 4.5.4 closely.

We refer to Theorems 2.1.4 and 2.1.5 for some consequences of the above theorem, if one also has
$$\int \phi^2 < \infty.$$

Example 4.7.2. If we put $\phi = \exp(-|x|^\lambda)$ where $\lambda > 0$ then the condition (4.7.3) becomes
$$|x|^\lambda \leqslant \varepsilon(\lambda^2|x|^{2\lambda-2} - \lambda(\lambda + N - 2)|x|^{\lambda-2}) - c_1 + c_2\varepsilon^{-\delta}$$
which holds for some $\delta > 0$ if and only if $\lambda > 2$. While this is an easy example, one should note that the corresponding potentials
$$V(x) = \lambda^2|x|^{2\lambda-2} - \lambda(\lambda + N - 2)|x|^{\lambda-2}$$
are rather special in form.

Our next theorem imposes hypotheses which correspond roughly to the potential V increasing no faster than quadratically at infinity.

Theorem 4.7.3. *Suppose that $0 < \phi \in C^2(\mathbb{R}^N) \cap L^2(\mathbb{R}^N)$ and that $V = \Delta\phi/\phi$ is bounded below. If there exist positive constants a and c such that*
$$\phi(x) \geqslant ce^{-ax^2}$$
for all $x \in \mathbb{R}^N$, then $C_0(\mathbb{R}^N)$ is invariant under $\exp(-H_\phi t)$ for all $t > 0$.

Proof. We see that ϕ is the ground state of e^{-Ht} and that the heat kernels are related by
$$K_\phi(t, x, y) = \phi(x)^{-1}\phi(y)^{-1}K(t, x, y).$$
Assuming that $V(x) \geqslant E$ for all $x \in \mathbb{R}^N$, a use of the Feynman–Kac formula shows that $K(t, x, y)$ is positive and continuous with
$$0 < K(t, x, y) \leqslant (4\pi t)^{-N/2}\exp\{-Et - (x-y^2)4t\}.$$

Therefore
$$0 < K_\phi(t, x, y) \leq c_t \exp\{ax^2 + ay^2 - (x-y)^2/4t\}.$$
Now
$$(x-y)^2 \geq \tfrac{1}{2}x^2 - y^2$$
so
$$0 < K_\phi(t, x, y) \leq c_t \exp\{(a - 1/8t)x^2 + (a + 1/4t)y^2\}.$$
We deduce that if $0 < t < (8a)^{-1}$ and $f \in C_c(\mathbb{R}^N)$ then
$$e^{-H_\phi t}f(x) = \int K_\phi(t, x, y) f(y) \phi^2(y) \, dy$$
vanishes as $|x| \to \infty$. Thus
$$\exp(-H_\phi t)(C_c) \subseteq C_0$$
for such t, and C_0 is invariant by the boundedness of $\exp(-H_\phi t)$ in the uniform norm. For $t \geq (8a)^{-1}$ the same result holds by the semigroup property.

Corollary 4.7.4. *Under the above hypotheses $\exp(-H_\phi t)$ cannot be ultracontractive.*

Proof. If $\exp(-H_\phi t)$ is ultracontractive then by Theorem 4.2.5 there exists T such that $t \geq T$ implies
$$K_\phi(t, x, y) \geq \tfrac{1}{2}$$
for all $x, y \in \mathbb{R}^N$. If $0 \leq f \in C_c(\mathbb{R}^N)$ it follows that
$$\exp(-H_\phi t) f(x) = \int K_\phi(t, x, y) f(y) \phi(y)^2 \, dy$$
$$\geq \tfrac{1}{2} \int f(y) \phi(y)^2 \, dy$$
for all $x \in \mathbb{R}^N$, which contradicts the conclusion of Theorem 4.7.3 that $\exp(-H_\phi t) f \in C_0(\mathbb{R}^N)$.

We now return to the general situation described by (4.7.1) and (4.7.2). For the remainder of this section we concentrate on the case where $\Omega = \mathbb{R}^N$ and there exist positive constants λ, μ and real constants α, β such that
$$\lambda^{-1}(1 + x^2)^\alpha \leq a(x) \leq \lambda(1 + x^2)^\alpha \qquad (4.7.4)$$
$$\mu^{-1}(1 + x^2)^\beta \leq b(x) \leq \mu(1 + x^2)^\beta \qquad (4.7.5)$$
for all $x \in \mathbb{R}^N$. We shall define L^p by the finiteness of the norm
$$\|f\|_p^p = \int_{\mathbb{R}^N} |f|^p b \, dx$$

for all $1 \leqslant p < \infty$, but also put

$$\|\|f\|\|_p^p = \int_{\mathbb{R}^N} |f|^p \, dx.$$

An important special case occurs when H is the Laplace–Beltrami operator for a (measurable) conformal Riemannian metric. This corresponds to the choice

$$a = b^{1-2/N}$$

and hence to

$$\alpha = \beta(1 - 2/N)$$

given (4.7.4) and (4.7.5). The Riemannian metric is then

$$ds = b(x)^{1/N}|dx|$$
$$\sim |x|^{2\beta/N}|dx|$$

as $|x| \to \infty$. One sees that $\mathbb{R}^N$ has finite volume for this metric if and only if $\beta < -N/2$, and that $\mathbb{R}^N$ is complete if and only if $\beta \geqslant -N/2$. However, if b is not smooth then the curvature of $\mathbb{R}^N$ cannot be defined pointwise.

It turns out that a number of spectral properties of H depend solely upon the values of α and β. We shall not attempt to give a complete analysis, but shall prove ultracontractivity for one range of values of α and β. Our main result is the following.

Theorem 4.7.5. *If $N \geqslant 3$ and*

$$\alpha \geqslant \beta(1 - 2/N) \tag{4.7.6}$$

then the heat kernel of e^{-Ht} satisfies

$$0 \leqslant K(t, x, y) \leqslant ct^{-N/2} \tag{4.7.7}$$

for all $0 < t \leqslant 1$ and $x, y \in \mathbb{R}^N$.

Lemma 4.7.6. *It is sufficient to prove (4.7.7) if*

$$\alpha = \beta(1 - 2/N). \tag{4.7.8}$$

Proof. If $\alpha > \beta(1 - 2/N)$ we put

$$a_1(x) = a(x)(1 + x^2)^{\alpha_1 - \alpha}$$

where

$$\alpha_1 = \beta(1 - 2/N).$$

Then $a_1(x)$ are the coefficients of a second order operator H_1 whose form Q_1 satisfies $Q_1 \leqslant Q$. If

$$0 \leqslant K_1(t, x, y) \leqslant c_1 t^{-N/2}$$

for all $0 < t \leq 1$ then Example 2.3.2 with $P = 0$ implies that

$$\int bf^2 \log f \leq \varepsilon Q_1(f) + \beta(\varepsilon)\|f\|_2^2 + \|f\|_2^2 \log \|f\|_2 \qquad (4.7.9)$$

for all $0 < \varepsilon \leq 1$ and $0 \leq f \in \operatorname{Quad}(H_1) \cap L^1 \cap L^\infty$, where

$$\beta(\varepsilon) = c - (N/4)\log \varepsilon.$$

The noted monotonicity now implies (4.7.9) with Q_1 replaced by Q, and the lemma follows by a second application of Example 2.3.2.

We henceforth assume (4.7.8).

Lemma 4.7.7. *There exists a constant $c_1 > 0$ such that the potential*

$$w(x) = (1 + x^2)^{-1 - 2\beta/N}$$

satisfies

$$w \leq c_1(H + 1).$$

Proof. If $\beta \geq -N/2$ then the result is trivial since w is bounded. If $\beta < -N/2$ then we put

$$\phi(x) = (1 + x^2)^\mu$$

where

$$\mu = \tfrac{1}{2}(1 - \alpha - N/2) \neq 0.$$

We have

$$\begin{aligned}H\phi &= -\mu b^{-1}\nabla \cdot \{\lambda^{-1}(1 + x^2)^{\alpha + \mu - 1} 2x\}\\ &= \mu b^{-1}\lambda^{-1}\{(1 - \mu - \alpha)(1 + x^2)^{\alpha + \mu - 2} 4x^2 - (1 + x^2)^{\alpha + \mu - 1} 2N\}\\ &= b^{-1}\lambda^{-1}(1 + x^2)^{\alpha + \mu - 2}(4\mu^2 x^2 - 2N\mu).\end{aligned}$$

Theorem 1.5.12 now implies that there exists $c_2 > 0$ such that

$$\begin{aligned}H &\geq b^{-1}\lambda^{-1}(1 + x^2)^{\alpha - 2}(4\mu^2 x^2 - 2N\mu)\\ &\geq c_2(1 + x^2)^{\alpha - \beta - 1}\frac{4\mu^2 x^2 - 2N\mu}{1 + x^2}.\end{aligned}$$

Therefore

$$4\mu^2 c_2(1 + x^2)^{\alpha - \beta - 1} \leq H + (2N\mu + 4\mu^2)(1 + x^2)^{\alpha - \beta - 2}$$

from which the lemma follows.

We next define a partition of $\mathbb{R}^N$ into cubes Ω_n as follows. For each integer $M \geq 1$ we divide $[-3^M, 3^M]^N$ into 3^N cubes each of edge length $2 \times 3^{N-1}$ and include in the partition all except the central cube. We then finally adjoin the cube $[-1, 1]^N$ and order the collection in terms of the integers $n \geq 1$. We let s_n be the edge length of Ω_n. It is not hard to see that there exist positive constants a_n, b_n, w_n and γ such that

$$\gamma^{-1}a_n \leqslant a(x) \leqslant \gamma a_n$$
$$\gamma^{-1}b_n \leqslant b(x) \leqslant \gamma b_n$$
$$\gamma^{-1}w_n \leqslant w(x) \leqslant \gamma w_n$$

for all $x \in \Omega_n$.

Proof of Theorem 4.7.5. Let K be the Laplacian with Neumann boundary conditions on the unit cube Ω, and K_n the corresponding operator on Ω_n. By applying Theorems 2.4.2 and 2.4.5, both with H replaced by $(K + 1)$, and also Lemma 1.7.11, we obtain

$$|V| \leqslant c|||V|||_{N/2}(K + 1)$$

for all potentials $V \in L^{N/2}(\Omega, dx)$. By scaling the independent variable x this implies

$$|V_n| \leqslant c|||V_n|||_{N/2}(K_n + s_n^{-2})$$

on each Ω_n, where V_n is the restriction to Ω_n of the potential V on $\mathbb{R}^N$. If $f \in C_c^\infty(\mathbb{R}^N)$ then

$$\langle |V_n| f, f \rangle = \int_{\Omega_n} |V_n| |f|^2 b \, dx$$

$$\leqslant \gamma b_n \int_{\Omega_n} |V_n| |f|^2 \, dx$$

$$\leqslant \gamma b_n |||V_n|||_{N/2} \int_{\Omega_n} (|\nabla f|^2 + s_n^{-2}|f|^2) \, dx$$

$$\leqslant \gamma^2 b_n a_n^{-1} |||V_n|||_{N/2} \int_{\Omega_n} |\nabla f|^2 a \, dx$$

$$+ \gamma^3 s_n^{-2} w_n^{-1} |||V_n|||_{N/2} \int_{\Omega_n} |f|^2 wb \, dx.$$

Now

$$(b_n a_n^{-1} |||V_n|||_{N/2})^{N/2} = b_n^{N/2} a_n^{-N/2} \int_{\Omega_n} |V_n|^{N/2} \, dx$$

$$\leqslant \gamma b_n^{N/2-1} a_n^{-N/2} \int_{\Omega_n} |V_n|^{N/2} b \, dx$$

$$\leqslant \gamma_1 \|V_n\|_{N/2}^{N/2}$$

where γ_1 is independent of n. Also

$$(s_n^{-2}w_n^{-1}|\!|\!|V_n|\!|\!|_{N/2})^{N/2} = s_n^{-N}w_n^{-N/2}\int_{\Omega_n}|V_n|^{N/2}\,dx$$

$$\leq \gamma s_n^{-N}w_n^{-N/2}b_n^{-1}\int_{\Omega_n}|V_n|^{N/2}b\,dx$$

$$\leq \gamma_2\|V_n\|_{N/2}^{N/2}$$

where γ_2 is independent of n. Therefore

$$\langle|V_n|f,f\rangle \leq \gamma_3\|V_n\|_{N/2}\int_{\Omega_n}|\nabla f|^2 a\,dx$$

$$+ \gamma_4\|V_n\|_{N/2}\int_{\Omega_n}|f|^2 wb\,dx.$$

Summing over n yields

$$\langle|V|f,f\rangle \leq \gamma_5\|V\|_{N/2}\int_{\mathbb{R}^N}(|\nabla f|^2 a + |f|^2 bw)\,dx$$

for all $f \in C_c^\infty(\mathbb{R}^N)$. Hence

$$|V| \leq \gamma_5\|V\|_{N/2}(H + w)$$
$$\leq \gamma_6\|V\|_{N/2}(H + 1)$$

by Lemma 4.7.7. The proof is completed by applying Theorems 2.4.2 and 2.4.5, both with H replaced by $(H + 1)$.

Note. For further discussion of the cases $N = 1, 2$ of Theorem 4.7.5, and of the cases where (4.7.6) fails, see the notes at the end of the chapter.

4.8 Potentials with local singularities

In this section we show how to apply all of our previous theory to operators of the form $(H_0 + V)$, where H_0 is a second order elliptic operator and V is a potential which may have local singularities. In our earlier theory, particularly Example 2.1.9 and Section 4.5, we have assumed that V is bounded below.

We start by assuming that $e^{-H_0 t}$ is a symmetric Markov semigroup on $L^2(\Omega)$, where Ω is a locally compact, second countable Hausdorff space. We assume that there exists a monotonically decreasing continuous function $\beta(\varepsilon)$ for $0 < \varepsilon < 1$, such that

$$\int f^2 \log f \leq \varepsilon Q_0(f) + \beta(\varepsilon)\|f\|_2^2 + \|f\|_2^2 \log \|f\|_2 \qquad (4.8.1)$$

for all $0 \leq f \in \text{Quad}(H_0) \cap L^1 \cap L^\infty$, and define $\mathscr{D}_+$ as in Lemma 2.2.6 by

$$\mathscr{D}_+ = \bigcup_{t>0} e^{-H_0 t}(L^1 \cap L^\infty)_+.$$

We also suppose that V is a potential on Ω which satisfies the quadratic form inequality

$$|V| \leq \delta H_0 + \gamma(\delta) \qquad (4.8.2)$$

for all $\delta > 0$ and some $\gamma(\delta)$. This implies that

$$(1-\delta)H_0 - \gamma(\delta) \leq H_0 + V \leq (1+\delta)H_0 + \gamma(\delta) \qquad (4.8.3)$$

for all $0 < \delta \leq 1$. Hence $(H_0 + V)$ may be defined as a self-adjoint operator by taking the form sum, and its form domain equals that of H_0.

Lemma 4.8.1. *Under the above assumptions we have the bound*

$$\int f^p \log f \leq \varepsilon \langle (H_0 + V)f, f^{p-1} \rangle + \Gamma(\varepsilon, p)\|f\|_p^p + \|f\|_p^p \log \|f\|_p$$

for all $f \in \mathscr{D}_+$, $0 < \varepsilon < 1$ and $2 \leq p < \infty$, where

$$\Gamma(\varepsilon, p) = (2/p)\beta(\varepsilon/2) + \varepsilon\gamma(1/p).$$

Proof. This is a variation upon the argument of Lemma 2.2.6. If $0 \leq g \in \text{Quad}(H_0) \cap L^1 \cap L^\infty$ then

$$\int g^2 \log g \leq (\varepsilon/2)Q_0(g) + \beta(\varepsilon/2)\|g\|_2^2 + \|g\|_2^2 \log \|g\|_2$$
$$+ (\varepsilon/2\delta)\{\langle Vg, g \rangle + \delta Q_0(g) + \gamma(\delta)\|g\|_2^2\}$$
$$= \varepsilon Q_0(g) + (\varepsilon/2\delta)\langle Vg, g \rangle$$
$$+ \{\beta(\varepsilon/2) + (\varepsilon/2\delta)\gamma(\delta)\}\|g\|_2^2 + \|g\|_2^2 \log \|g\|_2.$$

Putting $g = f^{p/2}$ and $\delta = p^{-1}$ we deduce from (2.2.7) that

$$\frac{p}{2}\int f^p \log f \leq \frac{\varepsilon p^2}{4(p-1)}\langle H_0 f, f^{p-1} \rangle + \frac{\varepsilon p}{2}\langle Vf, f^{p-1} \rangle$$
$$+ \frac{p}{2}\Gamma(\varepsilon, p)\|f\|_p^p + \frac{p}{2}\|f\|_p^p \log \|f\|_p$$

from which the Lemma follows immediately.

Theorem 4.8.2. *If (4.8.1) holds with*

$$\beta(\varepsilon) = v - (\mu/4)\log \varepsilon \qquad (4.8.4)$$

for some $v, \mu > 0$ and (4.8.2) holds with

$$\gamma(\delta) = \gamma_1 + \gamma_2 \delta^{-\alpha} \qquad (4.8.5)$$

for some $\gamma_i, \alpha > 0$, then
$$\|e^{-(H_0+V)t}f\|_\infty \leq at^{-\mu/4}\|f\|_2$$
for all $0 < t < 1$ and $f \in L^2$. Hence the heat kernel K of $e^{-(H_0+V)t}$ satisfies
$$0 \leq K(t,x,y) \leq a^2 t^{-\mu/2}$$
for $0 < t < 1$ and $x, y \in \Omega$.

Proof. We apply Theorem 2.2.7 with
$$\varepsilon(p) = 2^{2+\alpha}t(2+\alpha)p^{-2-\alpha}$$
which ensures that
$$t = \int_2^\infty p^{-1}\varepsilon(p)\,dp.$$
We first assume that V is bounded to simplify the domain problems. We then have
$$\|e^{-(H_0+V)t}\|_{\infty,2} \leq e^M$$
where
$$M = \int_2^\infty p^{-1}\Gamma(\varepsilon,p)\,dp$$
$$= 2\int_2^\infty p^{-2}\beta(2^{1+\alpha}t(2+\alpha)p^{-2-\alpha})\,dp$$
$$+ \gamma_1 \int_2^\infty 2^{2+\alpha}t(2+\alpha)p^{-2-\alpha}\,dp$$
$$+ \gamma_2 \int_2^\infty 2^{2+\alpha}t(2+\alpha)p^{-2}\,dp$$
$$= \gamma_3 + \gamma_4 t - (\mu/4)\log t.$$

If V is unbounded then we may approximate it by a sequence V_n of bounded potentials all of which satisfy the bound (4.8.2). One then has
$$\|e^{-(H_0+V_n)t}\|_{\infty,2} \leq at^{-\mu/2}$$
where a is independent of n, and the theorem follows by taking limits.

We now assume in addition to (4.8.1)–(4.8.5) that $e^{-H_0 t}$ is irreducible, that Ω has measure one and that $H_0 1 = 0$, so that Theorem 4.2.5 is applicable, with $\phi_0 = 1$.

Theorem 4.8.3. *Under the above hypothesis there exist a_t, b_t and $T > 0$ such that*
$$0 < a_t \leq K(t,x,y) \leq b_t < \infty \qquad (4.8.6)$$
for all $t \geq T$ and $x, y \in \Omega$.

Potentials with local singularities

Proof. The Trotter product formula implies that the heat kernel K_λ of $\exp\{-(H_0 + \lambda V)t\}$ is a logarithmically convex function of λ for every $t > 0$ and $x, y \in \Omega$. Hence

$$K_0(t, x, y)^2 \leqslant K_1(t, x, y) K_{-1}(t, x, y)$$

for all $t > 0$. Now Theorem 4.2.5 states that

$$\tfrac{1}{2} \leqslant K_0(t, x, y) \leqslant 2$$

if $t \geqslant T$, and Theorem 4.8.2 implies that

$$0 \leqslant K_{\pm 1}(t, x, y) \leqslant b_t$$

for all $t > 0$. Therefore

$$(4b_t)^{-1} \leqslant K_1(t, x, y)$$

as required.

Corollary 4.8.4. *If ϕ is the ground state of $(H_0 + V)$ normalised by $\|\phi\|_2 = 1$ then under the above conditions there exists $a > 0$ such that*

$$0 < a^{-1} \leqslant \phi(x) \leqslant a < \infty \tag{4.8.7}$$

for all $x \in \Omega$.

Proof. If the eigenvalue of $(H_0 + V)$ corresponding to ϕ is E, then integrating (4.8.6) against suitable factors yields

$$a_t \langle \phi, 1 \rangle^2 \leqslant e^{-Et} \leqslant b_t \langle \phi, 1 \rangle^2$$

and

$$a_t \langle \phi, 1 \rangle \leqslant e^{-Et} \phi \leqslant b_t \langle \phi, 1 \rangle$$

for all $t \geqslant T$. Therefore

$$e^{-2Et} \phi^2 \leqslant b_t^2 \langle \phi, 1 \rangle^2 \leqslant b_t^2 a_t^{-1} e^{-Et}$$

and

$$e^{-2Et} \phi^2 \geqslant a_t^2 \langle \phi, 1 \rangle^2 \geqslant a_t^2 b_t^{-1} e^{-Et}.$$

Combining these with the bound

$$|E| \leqslant \gamma(1)$$

deduced from (4.8.3), we obtain

$$a_t^2 b_t^{-1} e^{-\gamma(1)t} \leqslant \phi^2 \leqslant b_t^2 a_t^{-1} e^{\gamma(1)t}$$

for all $t \geqslant T$. This implies (4.8.7).

We now come to the application of these abstract bounds to elliptic operators. For the sake of simplicity we only consider the case $H_0 = -\Delta$ subject to Dirichlet boundary conditions on a bounded region Ω in $\mathbb{R}^N$, where $N \geqslant 3$.

Lemma 4.8.5. *If Ω is a bounded regular region in $\mathbb{R}^N$, and $V = V_1 + V_2$ where $V_1 \in L^p(\Omega)$ for some $p > N/2$, and where*

$$|V_2(x)| \leq a_1 + a_2 d(x)^{-\beta}$$

for some $0 < \beta < 2$ and all $x \in \Omega$, then V satisfies (4.8.2) for a $\gamma(\delta)$ of the form (4.8.5).

Proof. We treat the two parts of V separately, the first part being handled by Theorem 1.8.4. Secondly $\varepsilon > 0$ implies

$$|V_2| \leq a_1 + a_2 \{\varepsilon d^{-2} + (2\varepsilon/\beta)^{-\beta/(2-\beta)}\}$$

which leads to the required bound upon applying the regularity condition (1.5.5).

Theorem 4.8.6. *If Ω is a bounded regular region in $\mathbb{R}^N$ for $N \geq 3$, and V is as in Lemma 4.8.5, then the heat kernel K of $(H_0 + V)$ satisfies*

$$0 \leq K(t, x, y) \leq ct^{-N/2}$$

for all $x, y \in \Omega$ and $0 < t \leq 1$.

Proof. Example 2.1.8 implies that

$$\|e^{-H_0 t}\|_{\infty,2} \leq c_1 t^{-N/4}$$

for all $0 < t < \infty$, and Example 2.3.1 implies that (4.8.1) and (4.8.4) hold with $\mu = N$. We may now apply Theorem 4.8.2.

We comment that the behaviour of K for large t depends upon the value of the smallest eigenvalue E of $(H_0 + V)$. The main purpose of this section is to show that the behaviour of K near the boundary of Ω is exactly the same as that of the heat kernel K_0 of $e^{-H_0 t}$, and in particular that the two ground states ϕ and ϕ_0 are of the same order of magnitude. This does not make the boundary behaviour of ϕ trivial to analyse, but does at least mean that the potential V causes no extra complications to leading order.

Theorem 4.8.7. *Let Ω be a bounded region in $\mathbb{R}^N$ satisfying interior and exterior cone conditions, where $N \geq 3$. Let V be a potential satisfying the conditions of Lemma 4.8.5, and let $H_0 = -\Delta$ on $L^2(\Omega)$ subject to Dirichlet boundary conditions. If ϕ_0 is the ground state of H_0, then there exist $c > 0$ and $\mu > 0$ such that*

$$0 \leq K(t, x, y) \leq ct^{-\mu/2} \phi_0(x) \phi_0(y)$$

for all $0 < t \leq 1$ and $x, y \in \Omega$.

Proof. Let $U: L^2(\Omega, \phi_0^2 dx) \to L^2(\Omega, dx)$ be the unitary operator $Uf = \phi_0 f$

and let
$$\tilde{H}_0 = U^{-1}(H_0 - E_0)U$$
where E_0 is the eigenvalue of H_0 corresponding to ϕ_0. If $\tilde{K}_0$ and $\tilde{K}$ are the heat kernels of $e^{-\tilde{H}_0 t}$ and of $e^{-(\tilde{H}_0 + V)t}$ respectively then Theorem 4.6.9 implies that
$$0 \leq \tilde{K}_0(t, x, y) \leq c_1 t^{-\mu/2}$$
for $0 < t \leq 1$ and $x, y \in \Omega$. Theorem 4.8.2 now implies that
$$0 \leq \tilde{K}(t, x, y) \leq c_2 t^{-\mu/2}$$
for $0 < t \leq 1$ and $x, y \in \Omega$. Since
$$K(t, x, y) = e^{-E_0 t} \phi_0(x) \phi_0(y) \tilde{K}(t, x, y)$$
the theorem follows.

Corollary 4.8.8. *Under the above hypotheses there exist a_t, b_t and $T > 0$ such that $t \geq T$ implies*
$$0 < \tfrac{1}{2} a_t \leq \frac{K(t, x, y)}{K_0(t, x, y)} \leq 2 b_t < \infty \qquad (4.8.8)$$
for all $x, y \in \Omega$. Moreover, there exists a constant $a > 0$ such that
$$a^{-1} \phi_0(x) \leq \phi(x) \leq a \phi_0(x) \qquad (4.8.9)$$
for all $x \in \Omega$.

Proof. Theorem 4.8.3 states that
$$0 < a_t \leq \tilde{K}(t, x, y) \leq b_t < \infty$$
for large enough t, while Theorem 4.2.5 implies that
$$\tfrac{1}{2} \leq \tilde{K}_0(t, x, y) \leq 2$$
for large enough t. We deduce (4.8.8) upon using
$$\frac{\tilde{K}(t, x, y)}{\tilde{K}_0(t, x, y)} = \frac{K(t, x, y)}{K_0(t, x, y)}.$$
Secondly the ground state of $(\tilde{H}_0 + V)$ is $\tilde{\phi} = \phi/\phi_0$, and Corollary 4.8.4 states that
$$0 \leq a^{-1} \leq \tilde{\phi} \leq a < \infty$$
so (4.8.9) follows.

It is interesting that the above theorems may sometimes be used to obtain pointwise bounds on the ground state ϕ_0 of $-\Delta$. Our last theorem may be regarded as a generalisation of the subharmonic comparison theorem,

Proposition 4.4.4, which does not require the potential W of that theorem to be non-negative.

Theorem 4.8.9. *Suppose that $\Omega \subseteq \mathbb{R}^N$ is a bounded region in $\mathbb{R}^N$ where $N \geqslant 3$, and Ω satisfies interior and exterior cone conditions. Suppose that $0 < \phi \in C^2(\Omega) \cap W_0^{1,2}(\Omega)$ and that ϕ satisfies*
$$|\phi^{-1}\Delta\phi| = O(d^{-\alpha})$$
as $d \to 0$ for some $0 < \alpha < 2$. Then the ground state ϕ_0 of $-\Delta$ on Ω, subject to Dirichlet boundary conditions, satisfies
$$a^{-1}\phi \leqslant \phi_0 \leqslant a\phi$$
on Ω for some $a > 0$.

Proof. The hypothesis of the theorem states that
$$-\Delta\phi + V\phi = 0$$
on Ω, where $V = O(d^{-\alpha})$ as $d \to 0$. Thus V satisfies the conditions of Lemma 4.8.5 and we can apply Corollary 4.8.8.

Notes

Section 4.2 The idea of studying heat kernels by passing to the L^2 space weighted by the square of the ground state goes back to Nelson (1966) and Gross (1976), but the possibility of obtaining pointwise bounds in this context was first investigated by Davies and Simon (1984). Theorem 4.2.5 is taken from Davies (1986) and (1987C).

Section 4.3 Mehler's formula, given in Proposition 4.3.1, is well known; we refer to Davies (1980) p. 181 and Simon (1979) p. 38 for two among the many proofs. Theorem 4.3.4 is not the sharpest result of its kind. Nelson (1973) proved that e^{-Ht} is bounded from L^p to L^q if and only if
$$e^t \geqslant (q-1)^{\frac{1}{2}}(p-1)^{-\frac{1}{2}}$$
in which case e^{-Ht} is a contraction; see Gross (1976) for a proof based upon logarithmic Sobolev inequalities. Theorems 4.3.5 and 4.3.6 are taken from Davies and Simon (1986).

Section 4.4 The ideas behind Lemma 4.4.1 and Corollary 4.4.2 are due to Rosen (1976). Lemma 4.4.3 is taken from Davies (1985C). The subharmonic comparison theorem, Proposition 4.4.4, is classical in origin; see Davies

Potentials with local singularities

(1982) for a review of its applications to the spectral theory of Schrödinger operators.

Section 4.5 Lemma 4.5.1 and Theorem 4.5.4 were proved by Davies and Simon (1984) and other applications of the same ideas may be found in Davies (1985C). Theorem 4.5.10 is previously unpublished. Theorem 4.5.11 is due to Carmona (1974) and in a slightly more general form may be found in Davies and Simon (1984).

Section 4.6 For the Hopf boundary point lemma, Lemma 4.6.1, see Gilbarg and Trudinger (1977) and Davies and Simon (1984). Much more information about the boundary behaviour of eigenfunctions than that in Examples 4.6.5–4.6.7 may be found in Miller (1967; 1971), Oddson (1978) and Grisvard (1981/2). Theorems 4.6.8 and 4.6.9 are taken from Davies and Simon (1984) as refined by Davies (1987C). Theorem 4.6.11 was proved by Davies (1987C); for regions with C^2 boundaries, however, much more precise bounds on the Green function may be found in Hueber and Sieveking (1982), Hueber (1985) and Zhao (1986).

Section 4.7 Theorem 4.7.3 is extracted from Davies and Simon (1986). The method of proof of Theorem 4.7.5 is taken from Davies (1987A). A much more complete treatment for other values of α, β and N is given by Pang (1987); for other related spectral properties of these singular elliptic operators see Davies (1985B) and Pang (1987). See also Baouendi and Goulaouic (1969), Vulis and Solomjak (1972), Triebel (1978), Birman and Solomjak (1980), Davies and Mandouvalos (1987), Cordes (1987) and references there for results concerning other classes of singular elliptic operators.

Section 4.8 The theory of this section is taken from Davies (1986), but the class of potentials treated in Lemma 4.8.5 is influenced by Ancona (1986). We refer to Simon (1982) and Davies (1987B) for the proof of Gaussian upper bounds on the heat kernels of Schrödinger operators.

5
Riemannian manifolds

5.1 Fundamental properties of manifolds

The goal of this chapter is to obtain information about the heat kernel of a complete Riemannian manifold. There is a tremendous literature on this subject, much of which concerns the asymptotic form of the heat kernel $K(t, x, y)$ as $t \to 0$; we, however, shall be mainly interested in finding uniform upper and lower bounds over the whole range of t, x, y. For manifolds of non-negative Ricci curvature this problem is now largely solved as a result of the efforts of Li and Yau, whose work we shall describe below; see Corollary 5.3.6 and Theorems 5.5.6 and 5.6.3.

If one merely assumes that the Ricci curvature of the manifold is bounded below by a negative constant, the above methods can still be applied but give a much less complete picture. Indeed even for hyperbolic space and its quotients by Kleinian groups the variety of phenomena which can occur is vast and only partly understood. In Section 5.7 we give a brief summary of some recent results, without proofs.

We start with a brief introduction to manifold theory in order to fix notation. Let M be an n-dimensional (connected) manifold with tangent space TM and cotangent space T^*M. Smooth sections of TM are called vector fields and smooth sections of T^*M are called forms. Vector fields may alternatively be defined as maps $\xi: C^\infty(M) \to C^\infty(M)$ such that

$$\xi(fg) = f(\xi g) + g(\xi f)$$

for all $f, g \in C^\infty(M)$, and are given in a local coordinate system by

$$\xi = \sum_{i=1}^n \xi_i(x) \frac{\partial}{\partial x_i}.$$

If $f \in C^\infty(M)$ then the form df is defined by

$$\langle df, \xi \rangle = \xi f$$

using the natural pairing $T_x \times T_x^* \to \mathbb{R}$.

A Riemannian metric on M is defined as a smooth choice of a positive

Fundamental properties of manifolds

definite inner product $\langle \ , \ \rangle_x$ on T_x for each $x \in M$, so that
$$\langle \xi, \eta \rangle_x = \sum g_{ij}(x)\xi_i(x)\eta_j(x)$$
in local coordinates. The metric induces an isomorphism $j_x: T_x \to T_x^*$ and the corresponding inner product on T_x^* is given by
$$\langle \lambda, \mu \rangle_x = \sum g^{ij}(x)\lambda_i(x)\mu_j(x)$$
in local coordinates, where g^{ij} is the matrix inverse to g_{ij}. The induced distance function on $M \times M$ will be denoted by $d(x, y)$ and the natural integral for $f \in C_c^\infty(M)$ by
$$\int f = \int f(x)\,dx = \int f(x_1, \ldots, x_n) g^{\frac{1}{2}}(x)\,dx_1 \cdots dx_n$$
in local coordinates, where
$$g(x) = \det\{g_{ij}(x)\}.$$

If $f \in C^\infty(M)$ its gradient is defined by
$$\nabla f = j_x^{-1}\{df(x)\}$$
or in local coordinates by
$$(\nabla f)_i = g^{ij}\partial_j f$$
using the summation convention. If ξ is a vector field on M then its divergence $\nabla \cdot \xi$ is defined by the validity of
$$\int f \nabla \cdot \xi = -\int \nabla f \cdot \xi$$
for all $f \in C_c^\infty(M)$. Since
$$-\int \nabla f \cdot \xi = -\int g_{ik}\xi_k (g^{ij}\partial_j f) g^{\frac{1}{2}}\,dx_1 \cdots dx_n$$
$$= -\int \xi_j(\partial_j f) g^{\frac{1}{2}}\,dx_1 \cdots dx_n$$
$$= \int \partial_j(g^{\frac{1}{2}}\xi_j) f\,dx_1 \cdots dx_n$$
whenever f has support in a coordinate neighbourhood, we see that
$$\nabla \cdot \xi = g^{-\frac{1}{2}}\partial_j(g^{\frac{1}{2}}\xi_j).$$
The Laplacian is the operator on $C_c^\infty(M)$ defined by
$$\Delta f = \nabla \cdot (\nabla f)$$
$$= g^{-\frac{1}{2}}\partial_i(g^{\frac{1}{2}}g^{ij}\partial_j f)$$

in local coordinates. Since $f \in C_c^\infty$ implies

$$\langle -\Delta f, f \rangle = -\int \nabla \cdot (\nabla f) f$$

$$= \int |\nabla f|^2$$

$$\geq 0$$

we see that $-\Delta$ is a non-negative symmetric operator on $C_c^\infty(M)$ and that its form

$$Q(f) = \int |\nabla f|^2$$

is a Dirichlet form in the sense of the earlier chapters. Although we have concentrated up to now on elliptic operators defined on open regions in $\mathbb{R}^n$, much of what we have done generalises painlessly to manifolds and we shall not repeat the theory.

If we define H to be the form closure of $-\Delta$ then e^{-Ht} can be extended to a positivity-preserving one-parameter contraction semigroup on $L^p(M)$ for all $1 \leq p \leq \infty$ by Theorems 1.3.2, 1.3.3 and 1.3.5. Our task is to relate geometric assumptions about the manifold to properties of the semigroup and its heat kernel. We follow the convention of Section 1.4 in using the symbol H_p to denote the generator of e^{-Ht} in L^p when we want to emphasise its dependence upon p, so that $H = H_2$.

We conclude this section with some examples.

Example 5.1.1. Spherical polars in $\mathbb{R}^3$. We have

$$ds^2 = dr^2 + r^2 d\theta^2 + r^2 \sin^2\theta \, d\phi^2,$$

$$g_{ij} = \begin{pmatrix} 1 & & \\ & r^2 & \\ & & r^2 \sin^2\theta \end{pmatrix}, \quad g^{ij} = \begin{pmatrix} 1 & & \\ & r^{-2} & \\ & & r^{-2}\sin^{-2}\theta \end{pmatrix},$$

$$g = r^4 \sin^2\theta,$$

$$\int f = \int f(r, \theta, \phi) r^2 \sin\theta \, dr \, d\theta \, d\phi,$$

$$\Delta f = \frac{1}{r^2}\frac{\partial}{\partial r}\left(r^2 \frac{\partial f}{\partial r}\right) + \frac{1}{r^2 \sin\theta}\frac{\partial}{\partial \theta}\left(\sin\theta \frac{\partial f}{\partial \theta}\right) + \frac{1}{r^2 \sin^2\theta}\frac{\partial^2 f}{\partial \phi^2}.$$

Example 5.1.2. A conformal metric in two dimensions.

$$ds^2 = a(x, y)(dx^2 + dy^2)$$

where $a > 0$ on $\Omega \subseteq \mathbb{R}^2$. Then
$$g_{ij} = \delta_{ij}a, \quad g^{ij} = \delta_{ij}a^{-1},$$
$$\int f = \int f(x,y)a(x,y)\,dx\,dy,$$
$$\Delta f = a^{-1}(\partial^2 f/\partial x^2 + \partial^2 f/\partial y^2).$$

We see that the set of harmonic functions on Ω, that is the solutions of $\Delta f = 0$, is independent of the conformal factor a. This property does not extend to higher dimensions when one defines conformal metrics in a similar manner.

5.2 Regularity properties of the heat equation

We start with a classical result which is nevertheless highly non-trivial.

Theorem 5.2.1. *The semigroup* e^{-Ht} *on* $L^2(M)$ *has a strictly positive* C^∞ *kernel on* $(0, \infty) \times M \times M$.

Proof. If $f \in L^2(M)$ then $e^{-Ht}f$ lies in $\text{Dom}(H^n)$ for all $n \geq 1$ and all $t > 0$ by the spectral theorem. Local Sobolev embedding theorems now imply that $e^{-Ht}f$ is a C^∞ function of x for each $t > 0$. Since $t \to e^{-Ht}f$ is an analytic function of t with values in the Hilbert space $\text{Dom}(H^n)$ we can even conclude that $t, x \to e^{-Ht}f(x)$ is a C^∞ function on $(0, \infty) \times M$. For each $t > 0$ and $x \in M$ the map $f \to e^{-Ht}f(x)$ is bounded on L^2 so there exists $a(t,x) \in L^2$ with
$$e^{-Ht}f(x) = \langle f, a(t,x) \rangle.$$
The map $t, x \to a(t,x) \in L^2$ is weakly C^∞, and hence also norm C^∞ by Davies (1980) Section 1.5. If $g \in C_c^\infty$ then
$$\langle e^{-Ht}f, g \rangle = \int \langle f, a(t,x) \rangle g(x)\,dx$$
so we have the L^2 identity
$$e^{-Ht}g = \int a(t,x)g(x)\,dx.$$
If $g, h \in C_c^\infty$ then
$$\langle e^{-Ht}g, h \rangle = \langle e^{-Ht/2}g, e^{-Ht/2}h \rangle$$
$$= \int K(t,x,y)g(x)\overline{h(y)}\,dx\,dy$$

where
$$K(t, x, y) = \langle a(t/2, x), a(t/2, y) \rangle$$
is a C^∞ integral kernel.

Since e^{-Ht} is positivity-preserving we have $a(t, x) \geq 0$ for all t, x and hence also
$$K(t, x, y) \geq 0.$$
We have now only to prove that K is strictly positive.

If N is a regular coordinate neighbourhood in M and $K_0(t, x, y)$ is the heat kernel of $-\Delta$ in $L^2(N)$ with Dirichlet boundary conditions on ∂N then
$$0 < K_0(t, x, y) \leq K(t, x, y)$$
for $t > 0$ and $x, y \in N$ by Theorems 2.1.6 and 3.3.5. If $x, y \in M$ are arbitrary then by the connectedness of M there exists a chain $x = x_0, x_1, \ldots, x_{n+1} = y$ such that each pair x_{r-1}, x_r lie in a common coordinate neighbourhood N_r. Thus
$$K(t, x, y) \geq \int_{N_1 \times \cdots \times N_n} K(t/n, x, x_1)$$
$$\times K(t/n, x_1, x_2) \cdots K(t/n, x_n, y) \, dx_1 \cdots dx_n > 0.$$

If $\lambda > 0$ we shall write R_λ for the positivity-preserving operator on $(L^1 + L^\infty)$ which coincides with $(H + \lambda)^{-1}$ on L^p for all $1 \leq p \leq \infty$. If $f \in L^p$ then
$$R_\lambda f = \int_0^\infty e^{-\lambda t} T_t f \, dt$$
where this is interpreted in the strong sense for $1 \leq p < \infty$ and in the weak* sense for $p = \infty$. The integral kernel $G(\lambda, x, y)$ of R_λ, also known as the Green function, is correspondingly given by
$$G(\lambda, x, y) = \int_0^\infty e^{-\lambda t} K(t, x, y) \, dt$$
provided the integral converges, a matter which we shall investigate later.

Proposition 5.2.2. *If $f \in L^\infty$ and $\lambda > 0$ then $R_\lambda f$ is continuous and bounded. If $f \geq 0$ and f is not identically zero then $R_\lambda f$ is strictly positive.*

Proof. If $g = R_\lambda f$ then g is bounded and
$$\Delta g = \lambda g - f \in L^\infty$$
in the weak sense. Sobolev embedding theorems now imply that g is

Regularity properties of the heat equation

continuous. If $f \geq 0$ then

$$g(x) = \int_0^\infty \int_M e^{-\lambda t} K(t, x, y) f(y) \, dy \, dt$$

so $g(x) > 0$ by Theorem 5.2.1.

Theorem 5.2.3. *If M is a complete Riemannian manifold then $H = -\Delta$ is essentially self-adjoint on $C_c^\infty(M)$.*

Proof. Let $\psi: \mathbb{R} \to [0, 1]$ be a C^∞ function with $\psi(s) = 1$ if $0 \leq s \leq 1$ and $\psi(s) = 0$ if $2 \leq s < \infty$. Then define ϕ_n on M by

$$\phi_n(x) = \psi\{d(x, y)/n\}$$

where $y \in M$ is any fixed point. We see that $0 \leq \phi_n \leq 1$ and that ϕ_n are continuous functions of compact support which converge locally uniformly to 1 as $n \to \infty$. Moreover

$$\nabla \phi_n = \psi'(d/n) \nabla d / n$$

so

$$|\nabla \phi_n(x)| \leq \|\psi'\|_\infty / n$$

for all $x \in M$, and $\phi_n \in W_c^{1,2}(M)$.

Now suppose that H is not essentially self-adjoint on $C_c^\infty(M)$ and let $u \neq 0$ be a function in $L^2(M)$ orthogonal to $(H + 1)C_c^\infty$. Then u is a weak solution of

$$\Delta u = u$$

and hence is a C^∞ function by Sobolev embedding theorems. Also

$$0 \geq -\langle \phi_n^2 u, u \rangle$$

$$= \int \nabla(\phi_n^2 u) \cdot \nabla u$$

$$= \int \{\phi_n^2 (\nabla u)^2 + 2\phi_n u \nabla \phi_n \cdot \nabla u\}.$$

Putting

$$b_n(x) = |\nabla \phi_n|,$$
$$c_n(x) = \phi_n |\nabla u|,$$

we deduce that

$$\int c_n^2 \leq 2 \int \phi_n |u| |\nabla \phi_n| |\nabla u|$$

$$\leq 2 \int |u| b_n c_n.$$

Hence

$$\|c_n\|_2^2 \leq 2\|b_n\|_\infty \|u\|_2 \|c_n\|_2$$

and

$$\|c_n\|_2 \leq 2n^{-1}\|\psi'\|_\infty \|u\|_2$$
$$\to 0 \quad \text{as} \quad n \to \infty.$$

Fatou's lemma now implies that

$$\int |\nabla u|^2 = 0$$

so $\nabla u = 0$. We finally obtain the contradiction

$$u = \nabla \cdot (\nabla u) = 0$$

whether or not M has infinite volume.

The following technical lemma will be of critical importance below.

Lemma 5.2.4. *If $h \geq 0$ is continuous on M and $f \in L^\infty(M)$ and*

$$\lambda h - \Delta h \geq f \geq 0$$

for some $\lambda > 0$, then $0 \leq R_\lambda f \leq h$.

Proof. Let U_n be an increasing sequence of relatively compact open subsets of M with smooth boundaries and union equal to M. Then the self-adjoint operators K_n on $L^2(U_n)$ given by $K_n = -\Delta$ with Dirichlet boundary conditions satisfy $K_n \downarrow H$ in the sense of quadratic forms so

$$(K_n + \lambda)^{-1} \uparrow (H + \lambda)^{-1}$$

in the strong operator topology by Theorem 1.2.3. If χ_n denotes the characteristic function of U_n then

$$g_n = (K_n + \lambda)^{-1}(\chi_n f)$$

satisfy

$$\lambda g_n - \Delta g_n = f$$

on U_n. Therefore

$$\lambda(h - g_n) \geq \Delta(h - g_n)$$

in U_n and $(h - g_n) \geq 0$ on ∂U_n. The maximum principle now implies that $(h - g_n) \geq 0$ on U_n. If $m \leq n$ we deduce that

$$h \geq (\lambda + K_n)^{-1}(\chi_n f) \geq (\lambda + K_n)^{-1}(\chi_m f) \to (\lambda + H)^{-1}(\chi_m f)$$

as $n \to \infty$. Finally letting $m \to \infty$ we obtain

$$h \geq (\lambda + H)^{-1} f$$

as required.

For the remainder of this section we assume that M is a complete

Regularity properties of the heat equation

Riemannian manifold whose Ricci curvature satisfies
$$\text{Ric}(x) \geq -(n-1)a^2$$
for some $a \geq 0$ and all $x \in M$. The following theorem of Calabi is of fundamental importance.

Proposition 5.2.5. *If $\psi:[0,\infty) \to \mathbb{R}$ is C^2 and $y \in M$ then*
$$f(x) = \psi(d(x,y))$$
satisfies
$$\Delta f \leq \begin{cases} \psi''(d) + (n-1)a \coth(ad)\psi(d) & \text{if } a > 0 \\ \psi''(d) + (n-1)d^{-1}\psi(d) & \text{if } a = 0 \end{cases}$$
in the weak sense for all $x \neq y$.

Note. Because of problems with the cut locus one cannot assert that $x \to d(x,y)$ is a C^∞ function for $x \neq y$ unless x is close to y.

Theorem 5.2.6. *If M is complete with Ricci curvature bounded below then the following conservation of probability conditions hold:*

(i) $e^{-Ht}1 = 1$ *for all* $t \geq 0$,

(ii) $R_\lambda 1 = \lambda^{-1}1$ *for some (all)* $\lambda > 0$,

(iii) $C_c^\infty(M)$ *is a core of H_1 in $L^1(M)$.*

Proof. The equivalence of (i) and (ii) was proved in Theorem 1.4.4. We next prove (ii). Let $\psi(s) = 1 + s^2$ and
$$f(x) = \psi(d(x,y))$$
for some fixed $y \in M$. Then Proposition 5.2.5 implies
$$\Delta f \leq 2 + (n-1)a\coth(ad)f$$
$$\leq \{2 + (n-1)a\coth a\}f$$
provided $d(x,y) \geq 1$. It follows that there exists $\lambda > 0$ such that
$$\Delta f \leq \lambda f$$
for all $x \in M$. If $\varepsilon > 0$ and
$$g_\varepsilon = \lambda R_\lambda 1 - 1 + \varepsilon f$$
then
$$(\lambda - \Delta)g_\varepsilon = \lambda(\lambda - \Delta)R_\lambda 1 - \lambda + \varepsilon(\lambda f - \Delta f)$$
$$= \varepsilon(\lambda f - \Delta f)$$
$$\geq 0$$

on M. If
$$U_\varepsilon = \{x : \varepsilon f(x) < 1\}$$
then U_ε is a bounded set and
$$g_\varepsilon = \lambda R_\lambda 1 \geq 0$$
on ∂U_ε. The maximum principle now implies that $g_\varepsilon \geq 0$ on U_ε, or
$$1 \leq \varepsilon f(x) + \lambda R_\lambda 1(x)$$
for all $x \in U_\varepsilon$. Letting $\varepsilon \to 0$ we obtain
$$1 \leq \lambda R_\lambda 1$$
which implies (ii).

We finally prove (iii). If this is not true then there exists a non-zero $k \in L^\infty$ such that
$$\langle (1 + H_1)f, k \rangle = 0$$
for all $f \in C_c^\infty$, or equivalently
$$(1 - \Delta)k = 0$$
in the weak sense. We see that k is C^∞ by local elliptic regularity theorems. Putting
$$h = 1 \pm k/\|k\|_\infty$$
we obtain
$$\inf\{h(x) : x \in M\} = 0 \qquad (5.2.1)$$
and
$$h - \Delta h = 1.$$
Lemma 5.2.4 now implies that
$$h \geq R_\lambda 1 = 1$$
which contradicts (5.2.1).

Corollary 5.2.7. *If $f \in L^\infty(M)$ then $T_t f(x)$ is a bounded continuous function of (t, x) for all $t > 0$.*

Proof. If $f \geq 0$ then
$$T_t f(x) = \int K(t, x, y) f(y) \, dy$$
so Theorem 5.2.1 and Fatou's lemma imply that $T_t f(x)$ is lower semicontinuous. If $f \in L^\infty$ then
$$T_t f = T_t(f + \|f\|_\infty 1) - \|f\|_\infty 1$$
is again lower semicontinuous. Replacing f by $-f$ the result follows.

Regularity properties of the heat equation   155

The above corollary is one version of the Feller property for Markov semigroups. We next turn to a more restrictive version. We define $C_0(M)$ to be the space of continuous functions on M which vanish at ∞, and B to be the closed subspace of L^∞ on which T_t is strongly continuous. By Davies (1980) Section 1.4, B may be identified as the norm closure of the domain $\mathscr{D}$ of the generator H_∞. More precisely $f \in \mathscr{D}$ if and only if

$$H_\infty f = \lim_{t \to 0} -t^{-1}(T_t f - f)$$

exists as a weak* limit in L^∞.

Lemma 5.2.8. *We have $C_0(M) \subseteq B$ and $C_c^\infty(M) \subseteq \mathrm{Dom}(H_\infty)$ with*

$$H_\infty f = -\Delta f$$

for all $f \in C_c^\infty(M)$.

Proof. If $f, g \in C_c^\infty(M)$ then

$$\langle g, T_t f \rangle = \langle e^{-H_2 t} g, f \rangle$$

$$= \langle g, f \rangle + \int_0^t \langle e^{-H_2(t-s)} H_2 g, f \rangle \, ds$$

$$= \langle g, f \rangle - \int_0^t \langle g, T_{t-s} \Delta f \rangle \, ds.$$

Thus

$$|\langle g, T_t f - f \rangle| \leq \left| \int_0^t \langle g, T_{t-s} \Delta f \rangle \, ds \right|$$

$$\leq t \| g \|_1 \| \Delta f \|_\infty.$$

Since such g are dense in $L^1(M)$ we deduce that

$$\| T_t f - f \|_\infty \leq t \| \Delta f \|_\infty$$

so $f \in B$. Hence $C_0(M) \subseteq B$.

If $f \in C_c^\infty(M)$ then $\Delta f \in C_c^\infty(M) \subseteq B$ so $s \to T_{t-s} \Delta f$ is norm continuous. We deduce that

$$T_t f = f - \int_0^t T_{t-s}(\Delta f) \, ds$$

as a Banach space identity in L^∞. It follows that $f \in \mathrm{Dom}(H_\infty)$ with $H_\infty f = -\Delta f$.

Theorem 5.2.9. *If M is complete with Ricci curvature bounded below then:*

   (i) $T_t C_0(M) \subseteq C_0(M)$ *all $t > 0$,*

   (ii) $R_\lambda C_0(M) \subseteq C_0(M)$ *all $\lambda > 0$.*

Proof. In order to prove (ii) we put
$$f(x) = \psi(d(x, y))$$
where $y \in M$ is fixed and
$$\psi(s) = (1 + s^2)^{-1}.$$
Proposition 5.2.5 implies
$$\Delta f \leq O(d^{-4}) + (n-1)a \coth(ad)\psi(d)$$
as $d \to \infty$, whence there exists $\lambda > 0$ such that
$$\Delta f = \lambda f$$
for all $x \in M$. If $\mu > \lambda + 1$ we deduce that
$$\mu f - \Delta f \geq f$$
where $0 \leq f \in C_0(M)$. It follows from Lemma 5.2.4 that
$$0 \leq R_\mu f \leq f.$$
If $g \in C_c^\infty(M)$ then there exists $c < \infty$ such that
$$-cf \leq g \leq cf$$
and this implies
$$-cR_\mu f \leq R_\mu g \leq cR_\mu f$$
or
$$|R_\mu g| \leq cf.$$
Hence $g \in C_c^\infty(M)$ implies $R_\mu g \in C_0(M)$ for all $\mu > \lambda + 1$. We conclude that (ii) holds for all such μ. Using the formula
$$T_t h = \lim_{n \to \infty} \{(n/t)R_{n/t}\}^n h$$
valid for all $h \in B$, we now see that (i) holds for all $t > 0$. This implies (ii) for all $\lambda > 0$ by virtue of the formula
$$R_\lambda h = \int_0^\infty e^{-\lambda t} T_t h \, dt.$$

Although the converse of our next theorem is false, in many common situations a reverse implication does hold. For manifolds with exponential volume growth it is easy to obtain an upper bound to the bottom of the spectrum by the method below.

Theorem 5.2.10. *Let E be the bottom of the spectrum of $-\Delta$ acting on $L^2(M)$, where M is a complete Riemannian manifold. If the volume of M grows subexponentially then $E = 0$.*

The parabolic Harnack inequality

Proof. We first observe that for any $a \in M$

$$\inf \left\{ \frac{|B(a, r+1)| - |B(a, r)|}{|B(a, r)|} : 1 < r < \infty \right\} = 0. \tag{5.2.2}$$

For if the LHS is denoted λ, then

$$|B(a, r+1)| \geq (1 + \lambda)|B(a, r)|$$

so

$$|B(a, n)| \geq c(1 + \lambda)^n$$

for some $c > 0$ and all integers $n \geq 1$. But we are assuming that the volume grows subexponentially so $\lambda = 0$.

Next let $\phi_r \in W_c^{1,2}(M)$ be defined by

$$\phi_r(x) = \begin{cases} 1 & \text{if } d(x, a) \leq r \\ r + 1 - d(x, a) & \text{if } r \leq d(x, a) \leq r + 1, \\ 0 & \text{if } r + 1 \leq d(x, a). \end{cases}$$

Then

$$E \leq \int |\nabla \phi_r|^2 / \int |\phi_r|^2$$

$$\leq \frac{|B(a, r+1)| - |B(a, r)|}{|B(a, r)|}$$

and the theorem follows upon applying (5.2.2).

We conclude this section by stating without proof some properties of an important class of manifolds of bounded geometry.

Theorem 5.2.11. *Let Γ be a countable free group of isometries of a Riemannian manifold M and suppose that $\Gamma \backslash M$ is compact. Then 0 lies in the spectrum of the Laplacian on M if and only if Γ is amenable. Also M is transient, that is, has a finite Green function, unless Γ has a normal subgroup of finite index isomorphic to $\mathbb{Z}^d$ for $d = 0, 1$ or 2.*

5.3 The parabolic Harnack inequality

In this section we shall describe the approach of Li and Yau to proving a parabolic Harnack inequality which is much sharper than that in Corollary 3.3.6. Throughout the section we assume that M is a complete Riemannian manifold whose Ricci curvature is bounded below by $-K$. Our calculations are local in character and we assume that the functions we are examining are smooth.

Lemma 5.3.1. *If $f : M \to \mathbb{R}$ is smooth and the Ricci curvature tensor is*

bounded below by $-K$ then
$$\Delta\{(\nabla f)^2\} \geq (2/n)(\Delta f)^2 + 2\nabla(\Delta f)\cdot\nabla f - 2K(\nabla f)^2$$
where n is the dimension of M.

Proof. We verify the differential inequality at any one point a, which we choose to be the origin of a normal local coordinate system, so that
$$g^{ij} = \delta_{ij}, \qquad g_{ij} = \delta_{ij},$$
$$\Delta = \partial_{ii}, \qquad (\nabla f)^2 = (\partial_i f)(\partial_i f)$$
at a, where we have used the summation convention. In the coordinate neighbourhood we have
$$\partial_k g^{ij} = -g^{il}\Gamma^j_{lk} - g^{jl}\Gamma^i_{lk}.$$

At the point a we have
$$\Delta\{(\nabla f)^2\} = \partial_{ii}(g^{jk}\partial_j f \partial_k f)$$
$$= (\partial_{ii} g^{jk})\partial_j f \partial_k f + \partial_{ii}(\partial_j f \partial_j f)$$
$$= -\partial_i(g^{jl}\Gamma^k_{li} + g^{kl}\Gamma^j_{li})\partial_j f \partial_k f + 2\partial_{jii} f \partial_j f + 2\partial_{ij} f \partial_{ij} f$$
$$\geq -(\partial_i\Gamma^j_{ki} + \partial_i\Gamma^k_{ji})\partial_j f \partial_k f + 2\partial_{jii} f \partial_j f + (2/n)(\partial_{ii} f)^2$$
by the Schwarz inequality. Also
$$\nabla(\Delta f)\cdot\nabla f = \partial_j(g^{il}\partial_{il} f - g^{il}\Gamma^k_{il}\partial_k f)\partial_j f$$
$$= \partial_{jii} f \partial_j f - (\partial_j\Gamma^k_{ii})\partial_k f \partial_j f.$$

Therefore at a we have
$$\Delta\{(\nabla f)^2\} - 2\nabla(\Delta f)\cdot\nabla f \geq (2/n)(\Delta f)^2 + 2S^{jk}\partial_k f \partial_j f$$
where
$$S^{jk} = \partial_j\Gamma^k_{ii} - \tfrac{1}{2}\partial_i\Gamma^j_{ki} - \tfrac{1}{2}\partial_i\Gamma^k_{ji}.$$
This inequality is unaltered if we replace S^{jk} by the Ricci tensor
$$\tfrac{1}{2}(S^{jk} + S^{kj}) = R^{jk}$$
and we complete the proof by using the lower bound
$$R^{jk}\partial_k f \partial_j f \geq -K(\nabla f)^2.$$

Lemma 5.3.2. *Let $u > 0$ be a smooth solution of the heat equation*
$$\Delta u = u_t$$
on $M \times [0, T]$ and put
$$F(x, t) = t\{(\nabla f)^2 - \alpha f_t\} \tag{5.3.1}$$
where $f = \log u$ and $\alpha \in \mathbb{R}$. Then
$$\Delta F - F_t + 2\nabla f \cdot \nabla F + t^{-1}F \geq t[(2/n)\{(\nabla f)^2 - f_t\}^2 - 2K(\nabla f)^2]. \tag{5.3.2}$$

Proof. We first observe that f satisfies
$$\Delta f + (\nabla f)^2 = f_t$$
and that $G = tf_t$ satisfies
$$\begin{aligned}&\Delta G - G_t + 2\nabla f \cdot \nabla G + t^{-1}G \\ &= t\Delta f_t - f_t - t(\Delta f_t + 2\nabla f \cdot \nabla f_t) + 2t\nabla f \cdot \nabla f_t + f_t \\ &= 0.\end{aligned}$$

The LHS of (5.3.2) therefore equals
$$\begin{aligned}&t\Delta\{(\nabla f)^2\} - (\Delta f)^2 - 2t\nabla f \cdot \nabla f_t + 2t\nabla f \cdot \nabla\{(\nabla f)^2\} + (\nabla f)^2 \\ &\geqslant t[(2/n)(\Delta f)^2 + 2\nabla(\Delta f) \cdot \nabla f - 2K(\nabla f)^2 - 2\nabla f \cdot \nabla f_t + 2\nabla f \cdot \nabla\{(\nabla f)^2\}] \\ &= t[(2/n)(\Delta f)^2 - 2K(\nabla f)^2]\end{aligned}$$
which leads at once to the RHS of (5.3.2).

Lemma 5.3.3. *If M is compact and $\mathrm{Ric} \geqslant -K$ on M for some $K \geqslant 0$ then the function F defined by (5.3.1) satisfies*
$$F \leqslant \frac{n\alpha^2}{2}\left(1 + \frac{Kt}{2(\alpha - 1)}\right) \tag{5.3.3}$$
on $M \times [0, T]$ for any $\alpha > 1$.

Proof. Let (x, s) be the point in $M \times [0, t]$ at which F takes its maximum value. If $F(x, s) \leqslant 0$ then we have nothing to prove, while if $F(x, s) > 0$ then $s > 0$ so
$$\nabla F = 0, \qquad \Delta F \leqslant 0, \qquad F_s \geqslant 0$$
at (x, s). Lemma 5.3.2 now yields
$$s^{-2}F \geqslant (2/n)\{(\nabla f)^2 - f_s\}^2 - 2K(\nabla f)^2$$
at (x, s). If we put
$$\mu = F^{-1}(\nabla f)^2$$
then $\mu \geqslant 0$ and
$$F = s(F\mu - \alpha f_s)$$
so
$$f_s = \frac{F}{\alpha s}(\mu s - 1).$$
Therefore $\alpha > 1$ implies
$$s^{-2}F + 2K\mu F \geqslant \frac{2F^2}{n}\left(\mu - \frac{\mu s - 1}{\alpha s}\right)^2$$

Riemannian manifolds

and
$$F \leq \frac{n\alpha^2}{2} \frac{1 + 2K\mu s^2}{\{1 + (\alpha - 1)\mu s\}^2}$$
$$= \frac{n\alpha^2}{2} \frac{1 + Cv}{(1+v)^2}$$

where

and
$$v = (\alpha - 1)\mu s \geq 0$$

$$C = 2Ks/(\alpha - 1) > 0.$$

Finally $v \geq 0$ implies
$$1 + Cv \leq (1 + 2v + v^2)(1 + C/4)$$

so
$$F(x,t) \leq F(x,s) \leq \frac{n\alpha^2}{2}\left(1 + \frac{C}{4}\right) \leq \frac{n\alpha^2}{2}\left\{1 + \frac{Kt}{2(\alpha - 1)}\right\}$$

as stated in the lemma.

Lemma 5.3.4. *The same bound (5.3.3) holds if we assume that M is complete rather than compact.*

Proof. The problem is that we can no longer assume that F has any maximum $M \times [0, t]$. Let $a \in M$, $R > 0$ and let ϕ be a continuously differentiable function of compact support on M such that
$$0 \leq \phi(x) \leq \phi(a) = 1$$
for all $x \in M$ and
$$(\nabla \phi)^2 \leq \varepsilon \phi, \quad \Delta \phi \geq -\delta.$$

We shall construct such a function later.

Let (x, s) be the point in $M \times [0, t]$ at which ϕF takes its maximum value, and assume that this value is positive. Then
$$\nabla(\phi F) = 0, \qquad \Delta(\phi F) \leq 0, \qquad F_s \geq 0$$
at (x, s). Therefore
$$\phi \Delta F + 2\nabla \phi \cdot \nabla F \phi^{-1} + F\Delta \phi \leq 0.$$

Therefore
$$\phi \Delta F - 2F(\nabla \phi)^2 + F\Delta \phi \leq 0$$

and
$$\phi \Delta F \leq (2\varepsilon + \delta)F.$$

Lemma 5.3.2 implies that
$$\phi \Delta F - \phi F_t + 2\nabla f \cdot \nabla(\phi F) - 2F\nabla f \cdot \nabla \phi + t^{-1}\phi F$$
$$\geq t\phi[(2/n)\{(\nabla f)^2 - f_t\}^2 - 2K(\nabla f)^2]$$

The parabolic Harnack inequality

everywhere, so at (x, s) we have

$$F(2\varepsilon + \delta) - 2F\nabla f \cdot \nabla \phi + s^{-1}\phi F \geq s\phi[(2/n)\{(\nabla f)^2 - f_s\}^2 - 2K(\nabla f)^2].$$

Defining $u \geq 0$ as before, we obtain

$$F(2\varepsilon + \delta) + 2F^{\frac{3}{2}}u^{\frac{1}{2}}\varepsilon^{\frac{1}{2}}\phi^{\frac{1}{2}} + s^{-1}\phi F \geq s\phi\left[\frac{2F^2}{n\alpha^2 s^2}\{1 + (\alpha - 1)us\}^2 - 2KuF\right].$$

This implies

$$s(2\varepsilon + \delta) + 2sF^{\frac{1}{2}}u^{\frac{1}{2}}\varepsilon^{\frac{1}{2}}\phi^{\frac{1}{2}} + 1 + 2Kus^2 \geq \frac{2\phi F}{n\alpha^2}\{1 + (\alpha - 1)us\}^2.$$

This may be written in the form

$$A\lambda^2 - 2B\lambda - C \leq 0$$

where

$$\lambda = (\phi F)^{\frac{1}{2}},$$

$$A = \frac{2}{n\alpha^2}\{1 + (\alpha - 1)us\}^2,$$

$$B = su^{\frac{1}{2}}\varepsilon^{\frac{1}{2}},$$

$$C = 1 + 2Kus^2 + s(2\varepsilon + \delta).$$

We deduce that

$$\lambda \leq B/A + \{(B/A)^2 + C/A\}^{\frac{1}{2}}$$

and hence that

$$F(a, t) = \phi(a)F(a, t)$$
$$\leq \phi(x)F(x, s)$$
$$\leq \{B/A + ((B/A)^2 + C/A)^{\frac{1}{2}}\}^2.$$

Now

$$0 \leq \frac{B}{A} = \frac{n\alpha^2 u^{\frac{1}{2}}\varepsilon^{\frac{1}{2}}s}{2\{1 + (\alpha - 1)us\}^2}$$

$$\leq \frac{n\alpha^2 \varepsilon^{\frac{1}{2}} t^{\frac{1}{2}} u^{\frac{1}{2}} s^{\frac{1}{2}}}{2\{1 + (\alpha - 1)us\}^2}$$

$$\leq \frac{n\alpha^2 \varepsilon^{\frac{1}{2}} t^{\frac{1}{2}}}{2(\alpha - 1)^{\frac{1}{2}}}$$

which converges to zero as $\varepsilon \to 0$. Also

$$0 \leq \frac{C}{A} = \frac{n\alpha^2}{2}\left\{\frac{1 + 2Kus^2}{(1 + (\alpha - 1)us)^2} + \frac{s(2\varepsilon + \delta)}{(1 + (\alpha - 1)us)^2}\right\}$$

$$\leq \frac{n\alpha^2}{2}\left\{1 + \frac{Kt}{2(\alpha - 1)} + t(2\varepsilon + \delta)\right\}$$

which converges to the RHS of (5.3.3) as $\varepsilon \to 0$ and $\delta \to 0$. The proof of (5.3.3) now reduces to constructing a sequence ϕ_n with the stated properties for which $\varepsilon_n \to 0$ and $\delta_n \to 0$.

For this purpose we put
$$\phi_n(x) = \psi(d(x,a)^2/n)$$
where
$$\psi(\mu) = \begin{cases} (\mu-1)^2, & 0 \leq \mu \leq 1 \\ 0, & \mu \geq 1. \end{cases}$$

Then
$$\nabla\phi_n = \psi'\left(\frac{d^2}{n}\right)\frac{2d\nabla d}{n}$$
so
$$(\nabla\phi_n)^2 = \frac{4d^2}{n^2}\left\{\psi'\left(\frac{d^2}{n}\right)\right\}^2$$
$$\leq \frac{4}{n}(\psi')^2 \leq \frac{16}{n}.$$

Therefore
$$\varepsilon_n = 16/n \to 0 \quad \text{as} \quad n \to \infty.$$

Also
$$\Delta\phi_n = \psi''\left(\frac{d^2}{n}\right)\frac{4d^2(\nabla d)^2}{n^2} + \psi'\left(\frac{d^2}{n}\right)\frac{\Delta(d^2)}{n}$$
$$\geq -\left|\psi'\left(\frac{d^2}{n}\right)\right|\frac{c(d+1)}{n}$$
$$\geq -2c(n^{-\frac{1}{2}} + n^{-1})$$

by Proposition 5.2.5. Therefore
$$\delta_n = 2c(n^{-\frac{1}{2}} + n^{-1}) \to 0 \quad \text{as} \quad n \to \infty.$$

We can now state the main Harnack inequality of Li and Yau.

Theorem 5.3.5. *Let $u \geq 0$ be a solution of the heat equation on $M \times [0,T]$, where M is a complete Riemannian manifold with $\text{Ric} \geq -K$ for some $K \geq 0$. Then $0 < t < t + s \leq T$ and $\alpha > 1$ imply*
$$0 \leq u(x,t) \leq u(y, t+s)\left(\frac{t+s}{t}\right)^{n\alpha/2} \exp\left\{\frac{\alpha d^2}{4s} + \frac{n\alpha K s}{4(\alpha-1)}\right\}$$
where $d = d(x,y)$.

Proof. By replacing u by $(u + \varepsilon)$ and then letting $\varepsilon \to 0$ we see that we may assume in the proof that $u > 0$. Putting $f = \log u$ we see from Lemma 5.3.4

that
$$(\nabla f)^2 - \alpha f_t \leq \frac{n\alpha^2}{2t} + \frac{n\alpha^2 K}{4(\alpha-1)}.$$
Let
$$\beta = \frac{\alpha d^2}{4s^2} + \frac{nK\alpha}{4(\alpha-1)}$$
and
$$\phi(\lambda) = u(x_\lambda, \lambda)\lambda^{n\alpha/2} e^{\beta\lambda}$$
for $t \leq \lambda \leq t+s$, where $x_t = x$, $x_{t+s} = y$ and
$$|dx_\lambda/d\lambda| = d/s$$
so that x_λ moves along a geodesic between x and y. Then
$$\frac{d}{d\lambda}\log\phi = \nabla f \cdot \frac{dx}{d\lambda} + \frac{\partial f}{\partial \lambda} + \frac{n\alpha}{2\lambda} + \beta$$
$$\geq \nabla f \cdot \frac{dx}{d\lambda} + \frac{n\alpha}{2\lambda} + \beta + \alpha^{-1}(\nabla f)^2 - \frac{n\alpha}{2\lambda} - \frac{n\alpha K}{4(\alpha-1)}$$
$$\geq -|\nabla f|\frac{d}{s} + \frac{\alpha d^2}{4s^2} + \alpha^{-1}(\nabla f)^2$$
$$\geq 0.$$
Therefore ϕ is increasing and
$$u(x,t)t^{n\alpha/2}e^{\beta t} \leq u(y, t+s)(t+s)^{n\alpha/2} e^{\beta(t+s)}$$
as stated in the theorem.

Corollary 5.3.6. *If $u \geq 0$ is a solution of the heat equation on $M \times [0, T]$ where M is a complete manifold with $\text{Ric} \geq 0$, then $0 \leq t \leq t+s \leq T$ implies*
$$0 \leq u(x,t) \leq u(y, t+s)\left(\frac{t+s}{t}\right)^{n/2} e^{d^2/4s}.$$

Proof. We put $K = 0$ and then let $\alpha \to 1$.

Example 5.3.7. If in Euclidean space we put
$$u(x, t) = (4\pi t)^{-n/2} e^{-x^2/4t}$$
then
$$\frac{u(x,t)}{u(y, t+s)} = \left(\frac{t+s}{t}\right)^{n/2} \exp\left\{\frac{y^2}{4(t+s)} - \frac{x^2}{4t}\right\}.$$

Since
$$\sup\left\{\frac{y^2}{4(t+s)} - \frac{x^2}{4t} : |y-x| = d\right\} = \frac{d^2}{4s}$$
we see that Corollary 5.3.6 is sharp for this example.

Corollary 5.3.8. *If $u \geq 0$ is a solution of the heat equation on $M \times \mathbb{R}$, where M is a complete manifold with $\mathrm{Ric} \geq -K$ for some $K \geq 0$ then $t \in \mathbb{R}$ and $s > 0$ imply*
$$0 \leq u(x,t) \leq u(y, t+s) \exp\{(d + s(nK)^{\frac{1}{2}})^2/4s\}.$$

Proof. By letting the time origin move towards $-\infty$ we obtain
$$0 \leq u(x,t) \leq u(y, t+s) \exp\left\{\frac{\alpha d^2}{4s} + \frac{n\alpha K s}{4(\alpha - 1)}\right\}.$$
The exponential factor is minimised for
$$\alpha = 1 + (s/d)(nK)^{\frac{1}{2}}$$
and this proves the result.

Example 5.3.9. If we put
$$u_\alpha(x,t) = x_n^\alpha \exp\{-\alpha(n - 1 - \alpha)t\}$$
where $x \in \mathbb{H}^n$, the hyperbolic space of dimension n, then u_α is a solution of the heat equation, and
$$\frac{u_\alpha(x,t)}{u_\alpha(y, t+s)} = \left(\frac{x_n}{y_n}\right)^\alpha \exp\{\alpha(n - 1 - \alpha)s\}$$
Therefore
$$\sup\left\{\frac{u_\alpha(x,t)}{u_\alpha(y, t+s)} : d(x,y) = d\right\} = \exp\{\alpha d + \alpha(n - 1 - \alpha)s\}$$
and
$$\sup\left\{\frac{u_\alpha(x,t)}{u_\alpha(y, t+s)} : d(x,y) = d \text{ and } \alpha \in \mathbb{R}\right\} = \exp\{(d + (n-1)s)^2/4s\}.$$
This is very close, but not quite equal, to the bound of Corollary 5.3.8, since one has $\mathrm{Ric} = -(n-1)$ on $\mathbb{H}^n$.

5.4 Potential theory

Before continuing with our study of the heat equation we investigate some immediate consequences of (5.3.3) rewritten in the form
$$\frac{(\nabla u)^2}{u^2} - \alpha \frac{u_t}{u} \leq \frac{n\alpha^2}{2}\left\{\frac{1}{t} + \frac{K}{2(\alpha - 1)}\right\} \qquad (5.4.1)$$

Potential theory

where $u > 0$ is a solution of the heat equation on $M \times [0, T]$, M is a complete Riemannian manifold such that $\text{Ric} \geq -K$ for some $K \geq 0$, and $\alpha > 1$.

Theorem 5.4.1. *Let $v: M \to (0, \infty)$ be a solution of*
$$-\Delta v = \lambda v.$$
Then $\lambda \leq nK/4$ and
$$(\nabla v)^2/v^2 \leq nK/2 - \lambda + (nK/4 - \lambda)^{\frac{1}{2}}(nK)^{\frac{1}{2}}.$$

Proof. If we put
$$u(x, t) = v(x) e^{-\lambda t}$$
then u is a positive solution of the heat equation, so (5.4.1) yields
$$\frac{(\nabla v)^2}{v^2} + \lambda\alpha \leq \frac{n\alpha^2}{2}\left\{\frac{1}{t} + \frac{K}{2(\alpha - 1)}\right\}$$
for all $t > 0$ and $\alpha > 1$. Putting $\beta = \alpha - 1$ and letting $t \to +\infty$ we get
$$\frac{(\nabla v)^2}{v^2} \leq -\lambda(\beta + 1) + \frac{nK}{4}(\beta + 2 + \beta^{-1})$$
for all $\beta > 0$. The inequality $\lambda \leq nK/4$ is obtained by letting $\beta \to +\infty$, while the main inequality of the theorem is proved by choosing the β which minimises the RHS.

Corollary 5.4.2. *If v is a positive harmonic function on M then*
$$(\nabla v)^2 \leq nKv^2.$$
In particular if $\text{Ric} \geq 0$ on M then any positive or bounded harmonic function is constant.

Corollary 5.4.3. *If E is the bottom of the spectrum of H then*
$$0 \leq E \leq nK/4.$$
In particular if $\text{Ric} \geq 0$ on M then $E = 0$.

Proof. If $\lambda < E$ then there exists a function $u > 0$ on M such that
$$\Delta u = -\lambda u$$
Therefore
$$u(x, t) = u(x) e^{-\lambda t}$$
satisfies the heat equation on $M \times \mathbb{R}$ and Corollary 5.3.8 with $x = y$ implies
$$e^{\lambda s} \leq e^{nKs/4}$$

for all $s>0$. Hence $\lambda \leqslant nK/4$ for all $\lambda < E$, and the corollary follows. An alternative proof of the second statement may be obtained using Theorem 5.2.10 and Proposition 5.5.1.

We are now able to identify the Martin boundary (minimal positive solutions) of the heat equation on $M \times \mathbb{R}$. A minimal solution is a solution $u \geqslant 0$ such that if $0 \leqslant v \leqslant u$ and v is another solution then $v = \lambda u$ for some constant $\lambda \geqslant 0$.

Theorem 5.4.4. *If u is a minimal solution of the heat equation on $M \times \mathbb{R}$ where M is complete with Ricci curvature bounded below, then*

$$u(x,t) = v(x)\,e^{Et}$$

where $E \in \mathbb{R}$ and $v: M \to (0, \infty)$ satisfies

$$\Delta v = Ev.$$

Proof. If we put

$$w(x,t) = u(x, t-s)$$

where $s > 0$, then w is a solution of the heat equation and

$$0 \leqslant w(x,t) \leqslant u(x,t)\,e^{nKs/4}$$

by Corollary 5.3.8. Therefore

$$w(x,t) = \lambda_s u(x,t)$$

by the minimality of u. Assuming that u is not identically zero we see that $\lambda_s > 0$ and that

$$u(x, t-s) = \lambda_s u(x,t)$$

for all $s \in \mathbb{R}$, where $\lambda_{-s} = \lambda_s^{-1}$. Therefore

$$u(x,t) = v(x)\lambda_t^{-1}$$

where $v(x) = u(x, 0)$. The stated properties of $v(x)$ and λ_t follow by using the fact that u satisfies the heat equation.

Note 5.4.5. If there was no curvature restriction on M and the Brownian motion had a finite first passage time to infinity, then one could not expect Theorem 5.4.4 to be valid. The same happens if M is not complete.

We end this section with a simple but famous theorem of Lichnerowicz.

Theorem 5.4.6. *If M is a compact Riemannian manifold of dimension n with*

$$\text{Ric} \geqslant \rho > 0$$

then the smallest non-zero eigenvalue E of $-\Delta$ satisfies

$$E \geqslant n\rho/(n-1).$$

Proof. We integrate the inequality of Lemma 5.3.1 over M, where f is the eigenfunction of $-\Delta$ corresponding to the eigenvalue E. This yields

$$0 \geq (2/n)\int (\Delta f)^2 - 2E\int (\nabla f)^2 + 2\rho \int (\nabla f)^2$$

$$= \left(\frac{2E^2}{n} - 2E^2 + 2\rho E\right)\int f^2.$$

Since $E > 0$ and $\int f^2 \neq 0$ we deduce that

$$E/n - E + \rho \leq 0$$

which yields the stated lower bound on n.

If $\rho = 0$ then by considering spheres one sees that any lower bound on E must involve some further datum about M, such as its diameter or volume.

5.5 Upper bound on the heat kernel

Throughout this section we assume that M is a complete connected Riemannian manifold of dimension n with Ric ≥ 0. We do not, however, make any assumptions of bounded geometry. If $x \in M$ and $r > 0$ then $B(x, r)$ denotes the ball with centre x and radius r. For any set $E \subseteq M$, $|M|$ denotes its volume. If $H = -\Delta$ then the heat kernel $K(t, x, y)$ of e^{-Ht} is a positive C^∞ function by Theorem 5.2.1 and our goal is to obtain an upper bound for it in terms of geometrically computable quantities. We shall make fundamental use of the following geometric results.

Proposition 5.5.1. *Let M be complete with non-negative Ricci curvature. Let*

$$V(r) = |B(x, r)|$$

where $x \in M$ and $r > 0$, and let

$$V_0(r) = c_n r^n$$

be the volume of the Euclidean ball of radius r. Then

$$V(r)/V_0(r)$$

is a monotonically decreasing function of r. In particular:

(i) $\qquad V(r) \leq V_0(r)$

for all $r > 0$.

(ii) $\qquad V(r) \leq V(\alpha r) \leq \alpha^n V(r)$

for all $r > 0$ and $\alpha \geq 1$.

(iii) $\qquad \dfrac{V(t) - V(s)}{V(r)} \leq \dfrac{V_0(t) - V_0(s)}{V_0(r)}$

for all $0 < r \leq s \leq t$.

Analogous results exist for manifolds with Ricci curvature bounded below, but we shall not use them.

Theorem 5.5.2. *There exists a constant c depending only upon n such that*
$$0 < K(t, x, x) \leq c|B(x, t^{\frac{1}{2}})|^{-1}$$
for all $t > 0$ and $x \in M$.

Proof. Corollary 5.3.6 implies that
$$K(t, x, x) \leq K(t + s, y, x)\left(\frac{t+s}{t}\right)^{n/2} e^{d^2/4s}$$
for all $s > 0$ and $t > 0$ where $d = d(x, y)$. Integrating over $y \in B(x, r)$ yields
$$|B|K(t, x, x) \leq \left(\frac{t+s}{t}\right)^{n/2} e^{r^2/4s} \int_B K(t + s, y, x)\,dy$$
$$\leq \left(\frac{t+s}{t}\right)^{n/2} e^{r^2/4s}$$
since e^{-Ht} is a contraction semigroup on $L^1(M)$. The lemma follows upon putting $s = t = r^2$.

We next construct a suitably well-behaved function $\phi(x, t)$ which is of the same order of magnitude as $|B(x, t^{\frac{1}{2}})|^{-\frac{1}{2}}$. We start by defining the function $V(x, r)$ for $x \in M$ and $r > 0$ by
$$V(x, r) = \int_M f\{(d(x, y)/r\}\,dy$$
where f is a C^∞ decreasing function on $[0, \infty)$ such that $f(s) = 1$ if $0 \leq s \leq 1$ and $f(s) = 0$ if $s \geq 2$.

Lemma 5.5.3. *There exists a constant $c_1 > 0$ such that*
$$|B(x, r)| \leq V(x, r) \leq c_1 |B(x, r)|$$
for all $x \in M$ and $r > 0$.

Proof. It is immediate from the definition that
$$|B(x, r)| \leq V(x, r) \leq |B(x, 2r)|.$$
But Proposition 5.5.1 (ii) states that
$$|B(x, 2r)| \leq 2^n |B(x, r)|$$
and the lemma follows.

Lemma 5.5.4. There exist constants c_i such that
$$0 \leqslant \partial V/\partial r \leqslant c_2 V r^{-1}$$
$$|\nabla V| \leqslant c_3 V r^{-1}$$
$$-\Delta V \leqslant c_4 V r^{-2}$$
for all $x \in M$ and $r > 0$, where the derivatives are computed in the weak sense.

Proof. We have
$$\partial V/\partial r = r^{-1} \int_M \{-f'(d/r)d/r\} \, dy$$
where $f' \leqslant 0$. Therefore
$$0 \leqslant \partial V/\partial r \leqslant c_2' r^{-1} |B(x, 2r)|$$
$$\leqslant c_1 c_2' r^{-1} |B(x, r)|$$
$$\leqslant c_1 c_2' r^{-1} V(x, r).$$

Secondly
$$\nabla V = \int_M f'(d/r) \nabla d/r \, dy$$
so
$$|\nabla V| \leqslant \int_M |f'(d/r)| r^{-1} \, dy$$
$$\leqslant c_3' r^{-1} |B(x, 2r)|$$
$$\leqslant c_1 c_3' r^{-1} V(x, r).$$

Finally
$$\Delta V = \int_M f'(dr^{-1}) \Delta dr^{-1} \, dy + \int_M f''(dr^{-1}) |\nabla d|^2 r^2 \, dy.$$

Applying Proposition 5.2.5 with $a = 0$, we obtain
$$\Delta V \geqslant - \int_M |f'(dr^{-1})| \frac{n-1}{dr} \, dy - \int_M |f''(dr^{-1})| r^{-2} \, dy.$$

Hence
$$-\Delta V \leqslant \frac{n-1}{r^2} \int_M |f'(dr^{-1})| \frac{r}{d} \, dy + \frac{1}{r^2} \int_M |f''(dr^{-1})| \, dy$$
$$\leqslant c_4' r^{-2} |B(x, 2r)|$$
$$\leqslant c_1 c_4' r^{-2} V(x, r).$$

Lemma 5.5.5. If
$$\phi(x, t) = V(x, t^{\frac{1}{2}})^{-\frac{1}{2}}$$

then there exist positive constants c, A, F such that
$$c^{-1}|B(x,t^{\frac{1}{2}})|^{-\frac{1}{2}} \leq \phi(x,t) \leq c|B(x,t^{\frac{1}{2}})|^{-\frac{1}{2}}$$
$$0 \geq \partial\phi/\partial t \geq -(A/4t)\phi \tag{5.5.1}$$
$$\Delta\phi \leq (F/t)\phi \tag{5.5.2}$$
for all $x \in M$ and $t > 0$.

Proof. This is a direct computation using the bounds of the last two lemmas.

Theorem 5.5.6. *If M is complete with non-negative Ricci curvature then for all $\delta > 0$ there exists $c_\delta > 0$ such that*
$$0 < K(t,x,y) \leq c_\delta |B(x,t^{\frac{1}{2}})|^{-\frac{1}{2}} |B(y,t^{\frac{1}{2}})|^{-\frac{1}{2}}$$
$$\times \exp\{-d(x,y)^2/4(1+\delta)t\}$$
for all $t > 0$ and $x, y \in M$.

We shall, in fact, prove the equivalent bound
$$0 < K(t,x,y) \leq c_\delta \phi(x,t)\phi(y,t) \exp\{-d^2/4(1+\delta)t\} \tag{5.5.3}$$
by extending the arguments developed in Section 3.2. Note that Theorem 5.5.2 may be rewritten in the form
$$0 < K(t,x,x) \leq c\phi(x,t)^2. \tag{5.5.4}$$
Our argument is of an abstract character, depending only upon the bounds (5.5.1)–(5.5.4).

Given any $0 < T < \infty$ we define the unitary operator U_T from $L^2(M, \phi(x,T)^2 \, dx)$ to $L^2(M, dx)$ by
$$U_T f(x) = \phi(x,T) f(x)$$
and transfer the problem to this weighted space. Putting
$$d_T x = \phi(x,T)^2 \, dx$$
we define H_T on $L^2(M, d_T x)$ by
$$H_T = U_T^*(TH + F)U_T$$
so that H_T is associated with the form closure of
$$Q_T(f) = TQ(\phi_T f) + F\|f\|_2^2.$$
We define $\mathscr{B} = W_c^{1,\infty}$ to be the space of continuous functions of compact support on M such that ∇f calculated in the weak sense lies in L^∞. Since $\mathscr{B} \supseteq C_c^\infty(M)$ it is a form core for H. Since ϕ is a positive continuous function with locally bounded first derivatives $U_T^{\pm 1}\mathscr{B} = \mathscr{B}$, and $\mathscr{B}$ is also a form core

Upper bound on the heat kernel

of H_T. It will be of some significance that $f \in \mathscr{B}$ implies $|f|^\alpha \in \mathscr{B}_+$ for all $\alpha \geq 1$, and that $fg \in \mathscr{B}$ whenever $f \in \mathscr{B}$ and $g \in C^\infty$.

Now (5.5.2) implies that

$$Q_T(f) = T \int |\nabla(\phi_T f)|^2 \, dx + F \int |\phi_T f|^2 \, dx$$

$$= T \int \{|\nabla f|^2 + (F - T\phi_T^{-1}\Delta\phi_T)|f|^2\} \, d_T x$$

$$= T \int |\nabla f|^2 \, d_T x + \int |f|^2 \, d\mu_T \tag{5.5.5}$$

where μ_T is a non-negative measure. It follows as in Theorem 1.3.5 that $\exp(-H_T t)$ is a symmetric Markov semigroup on $L^2(M, d_T x)$; the measure $d\mu_T$ may be singular because of cut locus phenomena, but the second term on the RHS of (5.5.5) causes no problems.

Lemma 5.5.7. *The heat kernel K_T of $\exp(-H_T t)$ satisfies*

$$0 < K_T(t, x, y) \leq c_0 t^{-A/2}$$

for all $0 < t \leq 1$ and $x, y \in M$.

Proof. A direct calculation shows that

$$K_T(t, x, y) = e^{-Ft} \phi(x, T)^{-1} \phi(y, T)^{-1} K(tT, x, y). \tag{5.5.6}$$

Now K is the integral kernel of a positive operator so (5.5.4) implies

$$0 \leq K(s, x, y) \leq K(s, x, x)^{\frac{1}{2}} K(s, y, y)^{\frac{1}{2}}$$

$$\leq c\phi(x, s)\phi(y, s).$$

Therefore

$$0 < K_T(t, x, y) \leq c e^{-Ft} \frac{\phi(x, tT)}{\phi(x, T)} \frac{\phi(y, tT)}{\phi(y, T)}. \tag{5.5.7}$$

Also (5.5.1) implies that

$$\frac{\partial}{\partial t} \{t^{A/4} \phi(x, t)\} \geq 0$$

so

$$(tT)^{A/4} \phi(x, tT) \leq T^{A/4} \phi(x, T)$$

and

$$\phi(x, tT) \leq t^{-A/4} \phi(x, T).$$

The lemma follows by combining this with (5.5.7).

Lemma 5.5.8. *If $0 < \varepsilon < 1$ and $2 < p < \infty$ then*

$$\int f^p \log f \, d_T x \leq \varepsilon \langle H_T^{\frac{1}{2}} f, H_T^{\frac{1}{2}} f^{p-1} \rangle + 2\beta(\varepsilon) p^{-1} \|f\|_p^p + \|f\|_p^p \log \|f\|_p$$

for all $0 \leqslant f \in \mathcal{B}_+$, where

$$\beta(\varepsilon) = c_1 - \tfrac{1}{4}A \log \varepsilon.$$

Proof. This is a matter of combining Theorem 2.2.3 and Lemma 2.2.6.

We now repeat the arguments of Section 3.2. We put $\xi = e^{\alpha \psi}$ where $\alpha \in \mathbb{R}$ and $\psi: M \to \mathbb{R}$ is a bounded C^∞ function such that $|\nabla \psi| \leqslant 1$ everywhere on M.

Theorem 5.5.9. *If $\delta > 0$ and*

$$\lambda/(\lambda - 1) = 1 + \delta$$

then

$$0 < K_T(t, x, y) \leqslant c'_\delta t^{-A/2} \exp\{-d^2/4(1+\delta)tT\} \qquad (5.5.8)$$

for all $0 < t \leqslant \lambda^{-1}$ and $x, y \in M$.

Proof. We follow the same route as in Section 3.2. If $f \in \mathcal{B}_+$ and $2 \leqslant p < \infty$ then

$$\langle H_T^{\frac{1}{2}} f, H_T^{\frac{1}{2}}(f^{p-1}) \rangle \leqslant 2 \langle H_T^{\frac{1}{2}}(\xi f), H_T^{\frac{1}{2}}(\xi^{-1} f^{p-1}) \rangle + \alpha^2 p \|f\|_p^p.$$

The proof is as in Lemma 3.2.1, the extra term associated with the measure $\mu_T \geqslant 0$ only making matters easier. The proof now follows that of Theorem 3.2.5 except that when we put

$$\varepsilon(p) = \lambda 2^\lambda t p^{-\lambda}$$

we need to assume that $0 < t < \lambda^{-1}$ in order to ensure that $0 < \varepsilon(p) < 1$ for all $p \geqslant 2$.

Proof of Theorem 5.5.6. We put $t = \lambda^{-1}$ and $s = tT$ in (5.5.6) and (5.5.8) to obtain

$$K(s, x, y) = e^{F\lambda^{-1}} \phi(x, \lambda s) \phi(y, \lambda s) K_T(\lambda^{-1}, x, y)$$
$$\leqslant e^{F\lambda^{-1}} \lambda^{A/2} \phi(x, s) \phi(y, s) c'_\delta \lambda^{A/2} \exp\{-d^2/4(1+\delta)s\}.$$

The theorem follows by combining this with the bound on ϕ in Lemma 5.5.5.

We say that M has bounded geometry if there exists a function $v(r)$ and a constant $c > 0$ such that

$$c^{-1} v(r) \leqslant |B(x, r)| \leqslant c v(r)$$

for all $x \in M$ and $r > 0$; other notions of bounded geometry usually only imply this property for $0 < r \leqslant 1$.

Corollary 5.5.10. *If M is complete with non-negative Ricci curvature and*

bounded geometry then for all $\delta > 0$ there exists c_δ such that
$$0 < K(t, x, y) \leqslant c_\delta v(t^{\frac{1}{2}})^{-1} \exp\{-d^2/4(1+\delta)t\}$$
for all $t > 0$ and $x, y \in M$.

The following variation of Theorem 5.5.6 is sometimes useful.

Theorem 5.5.11. *If M is complete with non-negative Ricci curvature then for all $\delta > 0$ there exists $a_\delta > 0$ such that*
$$0 < K(t, x, y) \leqslant a_\delta |B(x, t^{\frac{1}{2}})|^{-1} \exp\{-d(x,y)^2/4(1+\delta)t\}$$
for all $t > 0$ and $x, y \in M$.

Proof. If $d = d(x, y)$ then Proposition 5.5.1 (ii) implies that
$$|B(x, t^{\frac{1}{2}})| \leqslant |B(y, d + t^{\frac{1}{2}})|$$
$$\leqslant |B(y, t^{\frac{1}{2}})| \left(\frac{d + t^{\frac{1}{2}}}{t^{\frac{1}{2}}}\right)^n$$
$$\leqslant b_\varepsilon^2 |B(y, t^{\frac{1}{2}})| e^{2\varepsilon d^2/t}$$
for any $\varepsilon > 0$. Therefore
$$0 < K(t, x, y) \leqslant c_\delta |B(x, t^{\frac{1}{2}})|^{-1} b_\varepsilon \exp\{\varepsilon d^2/t - d^2/4(1+\delta)t\}$$
by Theorem 5.5.6, for all $\varepsilon > 0$ and $\delta > 0$. The theorem follows.

5.6 Lower bounds on the heat kernel

We continue with the assumption that M is a complete Riemannian manifold of dimension n with Ric $\geqslant 0$.

Theorem 5.6.1. *We have*
$$K(t, x, y) \geqslant (4\pi t)^{-n/2} \exp\{-d(x,y)^2/4t\}$$
for all $x, y \in M$ and $t > 0$.

Proof. We apply Corollary 5.3.6 to the function
$$u(x, t) = K(t, x, y)$$
and obtain
$$K(s, x, x) \leqslant K(t + s, x, y) \left(\frac{t+s}{s}\right)^{n/2} e^{d^2/4t}$$
for all $s > 0$ and $t > 0$. Now local calculations establish that
$$1 = \lim_{s \to 0} (4\pi s)^{n/2} K(s, x, x)$$

so
$$1 \leqslant \liminf_{s \to 0} K(t+s, x, y)\{4\pi(t+s)\}^{n/2} e^{d^2/4t}$$
$$= K(t, x, y)(4\pi t)^{n/2} e^{d^2/4t}.$$

While this bound is extremely valuable because of the sharp constants, it is not of the same form as the upper bound in Theorem 5.5.6.

Lemma 5.6.2. *There exists a constant $a \geqslant 1$ such that*
$$\int_{d(x,y) \leqslant at^{\frac{1}{2}}} K(t, x, y)\, dy \geqslant \tfrac{1}{2}$$
for all $x \in M$ and $t > 0$.

Proof. Using the upper bound
$$K(t, x, y) \leqslant c|B(x, t^{\frac{1}{2}})|^{-1} \exp\{-d(x, y)^2/5t\}$$
of Theorem 5.5.11, we see that
$$\int_{d \geqslant at^{\frac{1}{2}}} K(t, x, y)\, dy \leqslant c|B(x, t^{\frac{1}{2}})|^{-1} \int_{d \geqslant at^{\frac{1}{2}}} e^{-d^2/5t}\, dy$$
$$= cV(t^{\frac{1}{2}})^{-1} \int_{r \geqslant at^{\frac{1}{2}}} e^{-r^2/5t} V(dr)$$
where
$$V(r) = |B(x, r)|.$$
By Proposition 5.5.1(iii) this is bounded above by
$$c_1 V_0(t^{\frac{1}{2}})^{-1} \int_{r \geqslant at^{\frac{1}{2}}} e^{-r^2/5t} V_0(dr)$$
$$= c_2 t^{-n/2} \int_{at^{\frac{1}{2}}}^{\infty} e^{-r^2/5t} nr^{n-1}\, dr$$
$$= c_2 \int_a^{\infty} e^{-s^2/5} ns^{n-1}\, ds$$
$$\leqslant \tfrac{1}{2}$$

for large enough $a \geqslant 1$. The proof is completed by noting that Theorem 5.2.6 implies that
$$\int_M K(t, x, y)\, dy = 1$$
for all $t > 0$ and $x \in M$.

Theorem 5.6.3. *Let M be a complete Riemannian manifold of dimension n with $\mathrm{Ric} \geqslant 0$. Then for all $0 < \delta < 1$ there exists $b_\delta > 0$ such that the heat*

kernel of M satisfies
$$K(t,x,y) \geq b_\delta |B(x,t^{\frac{1}{2}})|^{-1} \exp\{-d(x,y)^2/4(1-\delta)t\}$$
for all $t > 0$ and $x, y \in M$.

Proof. If a is the constant of Lemma 5.6.2 then $d(x,y) \leq at^{\frac{1}{2}}$ implies
$$K(t/2, x, y) \leq K(t,x,x) 2^{n/2} e^{d^2/2t}$$
$$\leq 2^{n/2} e^{a^2/2} K(t,x,x)$$
by Corollary 5.3.6. If B is the ball with centre x and radius $a(t/2)^{\frac{1}{2}}$ then Lemma 5.6.2 implies that
$$\tfrac{1}{2} \leq \int_B K(t/2, x, y)\,dy \leq c_1 |B| K(t,x,x).$$
An application of Proposition 5.5.1 (ii) now yields
$$c_2 |B(x, t^{\frac{1}{2}})|^{-1} \leq K(t,x,x)$$
for some $c_2 > 0$ and all $t > 0$ and $x \in M$.

If $\delta > 0$ then a second application of Corollary 5.3.6 yields
$$K(\delta t, x, x) \leq K(t, x, y) \delta^{-n/2} e^{d^2/4(1-\delta)t}.$$
Therefore
$$K(t,x,y) \geq \delta^{n/2} e^{-d^2/4(1-\delta)t} c_2 |B(x, (\delta t)^{\frac{1}{2}})|^{-1}$$
$$\geq b_\delta |B(x, t^{\frac{1}{2}})|^{-1} e^{-d^2/4(1-\delta)t}$$
where $b_\delta > 0$.

Note 5.6.4. By using the fact that
$$K(t,x,y) = K(t,y,x)$$
we can easily deduce the alternative lower bound
$$K(t,x,y) \geq b_\delta |B(x, t^{\frac{1}{2}})|^{-\frac{1}{2}} |B(y, t^{\frac{1}{2}})|^{-\frac{1}{2}} e^{-d^2/4(1-\delta)t}.$$

Corollary 5.6.5. *If M is a complete Riemannian manifold with non-negative Ricci curvature then*
$$\lim_{t \to 0} t \log K(t,x,y) = -d(x,y)^2/4$$
for all $x, y \in M$.

Proof. This follows immediately from Theorems 5.5.11 and 5.6.3 once one uses the fact that
$$B(x,r) \sim r^N$$
as $r \to 0$.

176   *Riemannian manifolds*

For extensions of this Corollary see the notes.

Theorem 5.6.6. *If M is complete with non-negative Ricci curvature then M is transient if and only if*

$$\int_1^\infty |B(x,t^{\frac{1}{2}})|^{-1}\,dt < \infty \tag{5.6.1}$$

for any (or all) $x \in M$.

Proof. It follows from the triangle inequality that if (5.6.1) is finite for some $x \in M$ then it is finite for all $x \in M$. The proof is now a simple matter of combining the definition of transience with Theorems 5.5.11 and 5.6.3.

5.7 Hyperbolic space

We describe some of the properties of hyperbolic space of dimension $(n+1)$ in the half-space model. We put

$$\mathbb{H}^{n+1} = \{z = (x,y) : x \in \mathbb{R}^n \quad \text{and} \quad 0 < y < \infty\}.$$

This is a complete Riemannian manifold for the conformal metric

$$ds^2 = y^{-2}(dx^2 + dy^2).$$

The volume element is

$$d\,\text{vol} = y^{-n-1}\,dx\,dy$$

where dx denotes the Euclidean volume element in $\mathbb{R}^n$. The Laplace–Beltrami operator is given by

$$\Delta = y^2(\Delta_x + \partial^2/\partial y^2) - (n-1)y(\partial/\partial y)$$

where Δ_x denotes the Euclidean Laplacian on $\mathbb{R}^n$.

The Riemannian distance $d = d(z_1, z_2)$ between two points $z_i \in \mathbb{H}^{n+1}$ may be computed from

$$\sigma = \tfrac{1}{2} + \tfrac{1}{2}\cosh d = \cosh^2(d/2)$$

where the function $\sigma = \sigma(z_1, z_2)$ is defined by

$$\sigma = \frac{|x_1 - x_2|^2 + (y_1 + y_2)^2}{4y_1 y_2}.$$

It follows from these last two formulae that the hyperbolic balls in $\mathbb{H}^{n+1}$ are also Euclidean balls; however, the hyperbolic and Euclidean centres do not coincide. We shall use $B(x,r)$ to denote the hyperbolic ball with hyperbolic centre x and hyperbolic radius r.

If we fix $a \in \mathbb{H}^{n+1}$ and put $\rho(z) = d(a,z)$ then it follows from the above

Hyperbolic space

formulae that
$$\Delta \rho = n \cosh \rho.$$

If f is a smooth function on $\mathbb{R}$ we deduce that
$$\Delta f(\rho) = f''(\rho) + n \coth \rho f'(\rho)$$
$$= (\operatorname{sh} \rho)^{-n} \frac{d}{d\rho} \left\{ (\operatorname{sh} \rho)^n \frac{df}{d\rho} \right\}. \tag{5.7.1}$$

The proof of a number of the above properties is most easily accomplished by using the symmetry properties of $\mathbb{H}^{n+1}$. If G denotes the group of all orientation-preserving isometries of $\mathbb{H}^{n+1}$, then G acts transitively on $\mathbb{H}^{n+1}$; up to central elements G can be identified with $SL(2, \mathbb{R})$ on $\mathbb{H}^2$ and with $SL(2, \mathbb{C})$ on $\mathbb{H}^3$. The group G contains the Euclidean isometry group of $\mathbb{R}^n$ in an obvious way, but is actually a simple Lie group of rank one. By using this symmetry group a number of problems can be reduced to a standard configuration, in which properties of $\mathbb{H}^2$ can be applied. Such methods enable one to show that the geodesics in $\mathbb{H}^{n+1}$ are semicircles with centres on the boundary $\mathbb{R}^n = \{(x, 0) : x \in \mathbb{R}^n\}$, and with ends which have tangents perpendicular to the boundary.

The function σ defined above satisfies
$$\sigma(gz_1, gz_2) = \sigma(z_1, z_2)$$
for all $g \in G$. It is called the fundamental point pair invariant because any function F on $\mathbb{H}^{n+1} \times \mathbb{H}^{n+1}$ which satisfies
$$F(gz_1, gz_2) = F(z_1, z_2)$$
for all $g \in G$ must be of the form
$$F = f(\sigma)$$
for some f. This applies in particular to the integral kernel of any operator which commutes with the action of G on $\mathbb{H}^{n+1}$.

The group symmetry implies that the Ricci curvature is constant on $\mathbb{H}^{n+1}$, and, in fact, a direct computation shows that
$$\operatorname{Ric} = -n.$$

It follows that all of the results in Section 5.2 and many of those in Section 5.3 are applicable to $\mathbb{H}^{n+1}$.

A direct computation shows that
$$-\Delta y^s = s(n-s) y^s$$
for all $s \in \mathbb{C}$. Upon putting $s = n/2$, it is immediate from Theorem 1.5.12 that
$$\operatorname{Sp}(-\Delta) \subseteq [n^2/4, \infty).$$

It turns out that the spectrum of $-\Delta_p$ acting on $L^p(\mathbb{H}^{n+1})$ depends upon p.

Theorem 5.7.1. *If $1 \leq p \leq \infty$ then the spectrum of $-\Delta$ acting on L^p is the part of the complex plane on or inside the parabola*

$$t \in \mathbb{R} \to n^2 p^{-1}(1 - p^{-1}) + t^2 + itn(1 - 2/p). \tag{5.7.2}$$

In particular the L^2 spectrum is the whole of the interval $[n^2/4, \infty)$.

We turn now to the heat kernel. Since the Laplacian commutes with the action of G on $\mathbb{H}^{n+1}$, it follows that

$$K(t, z_1, z_2) = K(t, gz_1, gz_2)$$

for all $g \in G$, $z_i \in \mathbb{H}^{n+1}$ and $t > 0$. Therefore

$$K(t, z_1, z_2) = k(t, d(z_1, z_2)).$$

From the fact that K is a solution of the heat equation and (5.7.1) we deduce that

$$\frac{\partial k}{\partial t} = \frac{\partial^2 k}{\partial \rho^2} + n \coth \rho \frac{\partial k}{\partial \rho}$$

for all $t > 0$ and $\rho > 0$. Moreover local analysis shows that

$$k(t, \rho) \sim (4\pi t)^{-(n+1)/2} e^{-\rho^2/4t}$$

for small ρ and t.

The heat kernel on $\mathbb{H}^2$ is of the form

$$k_2(t, \rho) = 2^{\frac{1}{2}}(4\pi t)^{-\frac{3}{2}} e^{-t/4} \int_\rho^\infty \frac{s e^{-s^2/4t}}{(\cosh s - \cosh \rho)^{\frac{1}{2}}} \, ds$$

while that on $\mathbb{H}^3$ is given by

$$k_3(t, \rho) = (4\pi t)^{-\frac{3}{2}} \frac{\rho}{\sinh \rho} e^{-t - \rho^2/4t}. \tag{5.7.3}$$

Other heat kernels may be calculated using Hadamard's method of descent, which yields the formula

$$k_n(t, \rho) = 2^{\frac{1}{2}} e^{(2n-1)t/4} \int_\rho^\infty \frac{\sinh \lambda \; k_{n+1}(t, \lambda)}{(\cosh \lambda - \cosh \rho)^{\frac{1}{2}}} \, d\lambda. \tag{5.7.4}$$

Iterating this once and then differentiating both sides leads to the formula

$$k_{n+2}(t, \rho) = -\frac{e^{-nt}}{2\pi \sinh \rho} \frac{\partial}{\partial \rho} k_n(t, \rho)$$

from which we can deduce, for example, that

$$k_5(t, \rho) = (4\pi t)^{-\frac{5}{2}} e^{-4t - \rho^2/4t} A(t, \rho)$$

where

$$A(t, \rho) = \left(\frac{\rho}{\sinh \rho}\right)^2 + 2t \frac{\rho \cosh \rho - \sinh \rho}{(\sinh \rho)^3}.$$

Hyperbolic space

The formulae in higher dimensions become increasingly complicated, so that it is not even obvious from them that the heat kernels are positive. Our next theorem establishes quantitative bounds on the heat kernel, and shows at the same time that one cannot expect a simple general theory for manifolds of negative curvature, such as we have developed earlier for manifolds of non-negative curvature.

Theorem 5.7.2. *For all $n \geq 1$ there exists a positive constant c_n such that*

$$c_n^{-1} h_{n+1}(t,\rho) \leq k_{n+1}(t,\rho) \leq c_n h_{n+1}(t,\rho)$$

for all $t > 0$ and $\rho > 0$, where

$$h_{n+1}(t,\rho) = (4\pi t)^{-(n+1)/2} \exp(-n^2 t/4 - n\rho/2 - \rho^2/4t)$$
$$\times (1+\rho+t)^{n/2-1}(1+\rho).$$

For $n = 2$ this is consistent with (5.7.3) because of the uniform equivalence

$$\rho/\sinh\rho \sim (1+\rho)e^{-\rho}.$$

If we fix $\rho > 0$ then the theorem implies that

$$k_{n+1}(t,\rho) \sim \begin{cases} t^{-(n+1)/2} & \text{as } t \to 0 \\ t^{-3/2} e^{-n^2 t/4} & \text{as } t \to \infty. \end{cases}$$

We next look at the large time behaviour of the diffusion associated with the heat kernel. The volume of the ball of radius r is independent of the centre of the ball, and is given by

$$|B(a,r)| = c_n \int_0^r (\sinh s)^n \, ds \qquad (5.7.5)$$

where c_n is the area of the unit sphere in the Euclidean space $\mathbb{R}^{n+1}$. One sees from this that

$$|B(a,r)| \sim e^{nr} \quad \text{as} \quad r \to \infty$$

a property which is closely associated with the negative curvature of $\mathbb{H}^{n+1}$; see Proposition 5.5.1 (ii).

Corollary 5.7.3. *If $\mu_t = e^{\Delta t}\delta_a$ then μ_t is a probability measure on $\mathbb{H}^{n+1}$ such that*

$$\lim_{t \to \infty} \mu_t\{z:(1-\varepsilon)nt \leq d(z,a) \leq (1+\varepsilon)nt\} = 1$$

for all $\varepsilon > 0$.

Note. Intuitively this states that the diffusion behaves like a motion with asymptotic velocity n directed away from the starting point. This behaviour is totally different from that of the Brownian motion in Euclidean space.

Proof. The fact that μ_t is a probability measure follows from Theorem 5.2.6. The group symmetry of $\mathbb{H}^{n+1}$ proves that the distribution of μ_t depends only upon the distance ρ from a. Theorem 5.7.2 and (5.7.5) together imply that

$$\mu_t(d\rho) = c_n(\sinh \rho)^n k_{n+1}(t, \rho)$$
$$\sim (1 + \rho^{-1})^{-n} t^{-(n+1)/2}(1 + \rho + t)^{n/2 - 1}(1 + \rho)$$
$$\times \exp(-n^2 t/4 + n\rho/2 - \rho^2/4t).$$

The dominant term in this is

$$\exp(-(\rho - nt)^2/4t)$$

which clearly concentrates around $\rho = nt$ for large t.

We now define a Kleinian group Γ to be a (countable) discrete group of isometries of $\mathbb{H}^{n+1}$. If Γ acts freely on $\mathbb{H}^{n+1}$ then $M = \Gamma \backslash \mathbb{H}^{n+1}$ is a manifold with Γ as fundamental group. A theorem of Selberg states that if Γ is countably generated then Γ contains a torsion-free normal subgroup Γ_0 of finite index, which therefore acts freely on $\mathbb{H}^{n+1}$. For this reason we confine attention below to Kleinian groups which act freely. Kleinian groups acting on $\mathbb{H}^2$ are called Fuchsian.

Example 5.7.4. The group $SL(2, \mathbb{Z})$ of all matrices $\begin{pmatrix} a & b \\ c & d \end{pmatrix}$ with integer entries and determinant one acts on $\mathbb{H}^2$ by

$$\begin{pmatrix} a & b \\ c & d \end{pmatrix} z = \frac{az + b}{cz + d}.$$

An application of the Euclidean algorithm shows that $SL(2, \mathbb{Z})$ is generated by γ_1 and γ_2 where $\gamma_1(z) = z + 1$ and $\gamma_2(z) = -z^{-1}$. The group does not act freely since the point i has fixed point group of order 2 generated by $\begin{pmatrix} 0 & 1 \\ -1 & 0 \end{pmatrix}$, and $\frac{1}{2} + \frac{\sqrt{3}}{2} i$ has fixed point group of order 3 generated by $\begin{pmatrix} 0 & 1 \\ -1 & 0 \end{pmatrix}$. If $\mathbb{Z}_2$ denotes the ring of integers mod 2, then $SL(2, \mathbb{Z}_2)$ has six elements and there is an obvious identification map

$$m: SL(2, \mathbb{Z}) \to SL(2, \mathbb{Z}_2).$$

The subgroup $\Gamma_0 = \ker m$ has index 6 in $SL(2, \mathbb{Z})$ and may be shown to act freely on $\mathbb{H}^2$.

If Γ is a Kleinian group which acts freely on H^{n+1} then $M = \Gamma \backslash \mathbb{H}^{n+1}$ inherits the metric of $\mathbb{H}^{n+1}$ and becomes a complete Riemannian manifold with constant negative curvature. M may be compact, but in Example 5.7.4, M is non-compact with finite volume.

A partial impression of the quotient manifold M can be obtained by considering a fundamental domain for the action of Γ on $\mathbb{H}^{n+1}$. If $a \in \mathbb{H}^{n+1}$ then we put

$$D_a = \{z \in \mathbb{H}^{n+1} : d(z,a) < d(z, \gamma a) \text{ all } e \neq \gamma \in \Gamma\}.$$

It may be seen that D_a is a convex polyhedron whose faces are parts of the geodesic surfaces

$$S_\gamma = \{z : d(z,a) = d(z, \gamma a)\}.$$

Under the quotient map $\pi : \mathbb{H}^{n+1} \to \Gamma \backslash \mathbb{H}^{n+1}$, D_a maps one-one onto a dense subset of M, but M differs from $\bar{D}_a$ in that there are various identifications on the boundary. Note that the quotient space M is canonical, while the group Γ has many fundamental domains other than those described above.

Since M is a complete Riemannian manifold of constant negative curvature, all of the theory of Section 5.2 is applicable and one can investigate the spectral properties of $-\Delta$ acting on $L^2(M)$. It turns out that this is an extraordinarily deep and rich problem involving ideas stretching all the way from number theory to ergodic theory and scattering theory. We can hardly begin to describe the ramifications, and shall just select a few topics which fit in with the main ideas of this book.

Example 5.7.5. We continue with Example 5.7.4. A fundamental domain for the action of $SL(2, \mathbb{Z})$ is the set

$$\{(x,y) : |x| < \tfrac{1}{2} \text{ and } x^2 + y^2 > 1\}$$

We see from this that the quotient space M has one cusp of infinite length but finite volume. However since $SL(2, \mathbb{Z})$ does not act freely, M is not a manifold, but is what is called an orbifold. If Γ_0 is the subgroup of index six described in Example 5.7.4 then a fundamental domain for its action is

$$\{(x,y) : |x| < 1 \text{ and } (x - \tfrac{1}{2})^2 + y^2 > \tfrac{1}{4} \text{ and } (x + \tfrac{1}{2})^2 + y^2 > \tfrac{1}{4}\}.$$

The quotient space M_0 is in this case a manifold of finite volume with three cusps. Moreover, M_0 is a six-fold covering of M, and has fundamental group Γ_0.

Example 5.7.6. The Hecke group Γ_a is defined for $a > 1$ by the two generators

$$\begin{pmatrix} 0 & 1 \\ -1 & 0 \end{pmatrix}, \begin{pmatrix} 1 & 2a \\ 0 & 1 \end{pmatrix}$$

and has fundamental domain

$$\{(x,y) : |x| < a \text{ and } x^2 + y^2 > 1\}.$$

Γ_a does not act freely and the quotient space has infinite volume. The bottom of the spectrum of $-\Delta$ acting on $L^2(\Gamma_a\backslash\mathbb{H}^2)$ is a positive eigenvalue E_a, and it is known that E_a is an analytic concave increasing function of a with $E_a \to 0$ as $a \to 1$ and $E_a \to \frac{1}{4}$ as $a \to \infty$.

Example 5.7.7. We define a Schottky domain to be the region in $\mathbb{H}^{n+1}$ exterior to a set $S_1,\ldots,S_m$ of mutually exterior spheres in $\mathbb{R}^{n+1}$ whose (Euclidean) centres lie on the boundary $\mathbb{R}^n$ of $\mathbb{H}^{n+1}$. We say that a Kleinian group is a Schottky group if it has a fundamental domain which is a Schottky domain. As an example if $S_1,\ldots,S_m$ are as described above and $\gamma_1,\ldots,\gamma_m$ are the inversions in the spheres, then the group Γ generated by $\gamma_1,\ldots,\gamma_m$ is a Schottky group.

If $n > 1$ then the quotient manifold M always has infinite volume if Γ_0 is a Schottky group. The bottom E_0 of the spectrum of $-\Delta$ acting on $L^2(M)$ varies continuously as the spheres S_i move without touching. If $n > 1$ then it is known that there is a constant $c_n > 0$ such that $E_0 \geq c_n$ for all Schottky groups acting on $\mathbb{H}^{n+1}$.

A central role in the spectral theory of Kleinian groups is played by the exponent of convergence $\delta(\Gamma)$. This is defined by

$$\delta(\Gamma) = \inf\{s > 0 : \sum_{\gamma\in\Gamma} \exp\{-sd(z,\gamma w)\} < \infty\}.$$

It follows easily from the triangle inequality that the convergence of the above series does not depend upon the choice of z and w in $\mathbb{H}^{n+1}$. We let $E_0(\Gamma)$ denote the bottom of the spectrum of $-\Delta$ acting on $L^2(M)$.

Theorem 5.7.8. *One always has*

$$0 \leq \delta(\Gamma) \leq n.$$

If M has finite volume then $\delta(\Gamma) = n$. If $\delta(\Gamma) \leq n/2$ then $E_0(\Gamma) = n^2/4$. If $n/2 < \delta(\Gamma)$ then $E_0(\Gamma) = \delta(n-\delta)$.

Note that if the volume of M grows subexponentially then $0 \in \text{Sp}(-\Delta)$ by Theorem 5.2.10, so one must have $\delta = n$. In order to obtain more detailed results, we restrict the class of Kleinian groups under study. We say that Γ is geometrically finite if it has a fundamental domain which is a polyhedron with a finite number of faces. All of the examples above are of this kind.

Theorem 5.7.9. *Let Γ be a geometrically finite Kleinian group with $n/2 < \delta(\Gamma) < n$. Then the spectrum of $-\Delta$ acting on $L^2(M)$ is of the form*

$$\{E_0,\ldots,E_k\} \cup [n^2/4, \infty)$$

where

$$0 < \delta(n-\delta) = E_0 < E_1 < \cdots < E_k < n^2/4$$

are L^2 eigenvalues of finite multiplicity, and E_0 has multiplicity one. Moreover, the part $[n^2/4, \infty)$ of the spectrum is purely absolutely continuous. If $1 \leq p \leq \infty$ then the L^p spectrum of $-\Delta$ consists of all points on or inside the parabola (5.7.2), together with all E_i such that

$$E_i < n^2 p^{-1}(1 - p^{-1}).$$

If Γ is geometrically finite with $0 \leq \delta \leq n/2$ then the L^2 spectrum is $[n^2/4, \infty)$ and is purely absolutely continuous, while the L^p spectrum consists of all points on or inside the parabola (5.7.2).

One of the ways of proving such results is by studying the heat kernel $\tilde{K}$ of M. This is given by the formula

$$\tilde{K}(t, \pi z, \pi w) = \sum_{\gamma \in \Gamma} K(t, z, \gamma w)$$

for all $t > 0$ and $z, w \in \mathbb{H}^{n+1}$, π being the canonical map of $\mathbb{H}^{n+1}$ onto M. It is clear from this formula that pointwise bounds on $\tilde{K}$ will require very precise control of K, such as is given in Theorem 5.7.2. It is also possible to obtain many results by analysing the Green function using potential theoretic results.

If Γ is a Kleinian group then its action on $\mathbb{H}^{n+1}$ induces an action on $\partial \mathbb{H}^{n+1} = \mathbb{R}^n \cup \{\infty\}$. The ordinary set $\Omega(\Gamma)$ of this action is defined to be the largest open set on which Γ acts discontinuously, and the limit set $L(\Gamma)$ is its complement; Γ clearly also acts on $L(\Gamma)$.

Theorem 5.7.10. *Let Γ be geometrically finite with $0 < \delta(\Gamma) < n$. Then the limit set $L(\Gamma)$ has Hausdorff dimension $\delta(\Gamma)$. There exists a unique probability measure μ on $L(\Gamma)$ such that*

$$(\gamma \cdot \mu)(dx) = j(\gamma, x)^{-\delta_\Gamma} \mu(dx)$$

where

$$j(\gamma, x) = \text{Im}\,(\gamma \cdot x)/\text{Im}\,(x).$$

The Patterson measure μ defined above is of great importance in the further theory of Kleinian groups, and brings in many notions from ergodic theory.

Notes

Section 5.1 We have not attempted to give a self-contained treatment of Riemannian manifolds, but simply specify the notation we are using. For other accounts from a similar point of view see Chavel (1984) and Cycon et al. (1987).

Section 5.2. Theorem 5.2.1 is a classical result. There are many existing proofs of Theorem 5.2.3; see Chernoff (1973), Davies (1985D) and Strichartz (1983) for further references. A proof of Theorem 5.2.3, due to Calabi (1965), may also be found in Chavel (1984) p. 185. Criteria for the conservation of probability go back to Hasminskii (1960), and a version of Theorem 5.2.6 applicable to elliptic operators may be found in Davies (1985D) or Stroock and Varadhan (1979) p. 254. Varopoulos (1983) has given sharp conditions on the rate at which the curvature can decrease at infinity, if one is to avoid explosions. A more general version of Theorem 5.2.9 is given by Azencott (1974) and may also be found in Davies (1985D). Theorem 5.2.10 is due to Brooks (1981A), who also has a number of further results. The amenability statement in Theorem 5.2.11 is due to Brooks (1981B), (1985). Lyons and Sullivan (1984) and Varopoulos (1984) showed that M is transient if and only if the random walk on Γ is transient. The classification of transient discrete groups was achieved by Varopoulos (1986) following earlier work in Varopoulos (1985A; 1985B).

Section 5.3 Most of the material is taken from Li and Yau (1986), which contains far more than our selection of results indicates. Lemma 5.3.1 is essentially the Bochner–Lichnerowicz–Weitzenbock formula.

Section 5.4 Corollary 5.4.2 should be compared with the bound

$$(\nabla v)^2 \leqslant (n-1)Kv^2$$

proved by Yau (1975). The existence of positive or bounded harmonic functions when the Ricci curvature of M is allowed to be negative on a compact subset is studied by Donnelly (1986) and Li and Tam (1987); more general results may be found in Lyons (1987) and Lyons and Sullivan (1984). The Martin boundary of negatively curved manifolds was determined by Anderson and Schoen (1985); see also Ancona (1987) and Kifer (1986). More detailed bounds on the bottom of the spectrum than those of Corollary 5.4.3 may be found in Brooks (1981A). Theorem 5.4.4 is taken from Koranyi and Taylor (1985); for other work in the same spirit see Kaufman and Wu (1982), Koranyi and Taylor (1983), Mair and Taylor (1984) Pinchover (1986; 1987). Theorem 5.4.6 is due to Lichnerowicz (1958). For a selection from the enormous literature concerning estimates of the smallest few eigenvalues of a compact Riemannian manifold see Li and Yau (1980), Donnelly and Li (1982), Chavel (1984) and Berard (1986).

Section 5.5 Proposition 5.5.1 is taken from Cheeger, Gromov and Taylor (1982), where a number of references to earlier work on such bounds may be found. Theorems 5.5.2 and 5.5.6 are due to Li and Yau (1986) and improve upon earlier work, surveyed in Chavel (1984), in not making any assumption of bounded geometry. Our proof of Theorem 5.5.6, however, is taken from Davies (1988B). Li and Yau also obtained heat kernel bounds for manifolds whose Ricci curvature is bounded below by a negative constant. These bounds are not so sharp, and have been improved for large times by Davies (1988B).

Section 5.6 The lower bound given in Theorem 5.6.1 was proved by Cheeger and Yau (1981), following earlier work of Debiard, Gaveau and Mazet (1976); they, in fact, gave a more general version applicable to manifolds with Ricci curvature bounded below. Our proof, however, follows Li and Yau (1986). Our proof of Theorem 5.6.3 is a variation upon that of Li and Yau (1986). Corollary 5.6.5 was proved by Varadhan (1967) for complete manifolds whose Ricci curvature is bounded below by a negative constant; our method may also be adapted to this case by using Li and Yau (1986). For more information about the large time behaviour of the heat kernel when $\text{Ric} \geq 0$ see Li (1986). Theorem 5.6.6 is due to Varopoulos (1981; 1982; 1983); see also Li and Yau (1986) and Li and Tam (1987; 1988). Varopoulos (1983) describes an example of R. Greene which shows that Theorem 5.6.6 fails for general complete manifolds of bounded geometry.

Section 5.7 A comprehensive account of the geometry of hyperbolic space is given by Beardon (1983). The structure of the spectrum of $-\Delta$ acting on $L^2(H^{n+1})$ is classical, but the L^p spectrum was determined by Lohoué and Rychener (1978; 1982); see also Anker and Lohoué (1986) for an analogous result on a general class of symmetric spaces of rank greater than one. Hadamard's method of descent is described in Courant and Hilbert (1966) for Euclidean space. Its use in hyperbolic space appears to be due to Millson, and formulae such as (5.7.4) may be found in Debiard, Gaveau and Mazet (1976), Lohoué and Rychener (1978; 1982), Cheeger and Taylor (1982), Lax and Phillips (1982), Bougerol (1983), Davies and Mandouvalos (1988). Theorem 5.7.2 is due to Davies and Mandouvalos (1988) but its Corollary 5.7.3 has been folk knowledge for some years; for another approach to these problems see Lyons (1988). Selberg's theorem on torsion-free subgroups may be found in Cassels (1986) p. 87. Example 5.7.4 is the classical example of a Fuchsian group, but its spectral theory is nevertheless extremely deep; see Terras (1985) for a fascinating review. The spectral properties of the Hecke group given in Example 5.7.6 were proved by Phillips and Sarnak (1985A; 1985B). The existence of a positive lower bound to $E_0(\Gamma)$ for Schottky groups was proved by Phillips and Sarnak (1985B) in dimension greater than three, and by Doyle (1988) in dimension three, following earlier work of Bridges. Theorems 5.7.8 and 5.7.10 were proved for geometrically finite Fuchsian groups by Elstrodt (1974) and Patterson (1976) and then in greater generality by Sullivan (1979; 1984; 1987). The L^2 part of Theorem 5.7.9 is due to Lax and Phillips (1982; 1985) and the L^p part to Davies, Simon and Taylor (1988). A detailed analysis of the continuous spectrum has been given by Mandouvalos (1983; 1986; 1987; 1988) by developing a theory of Eisenstein series. The relationship between Eisenstein series and scattering theory has been investigated by Mandouvalos (1987) and Perry (1987A; 1987B). Theorem 5.7.10 fails for general Kleinian groups; indeed Patterson (1983) has shown that the exponent of convergence may be arbitrarily small even though the limit set is the whole of $\mathbb{R}^n \cup \{\infty\}$. For more detailed accounts of these topics see Patterson (1987) and Sullivan (1987).

References

Adams, R. A. (1975) *Sobolev spaces*. Academic Press.
Agmon, S. (1982) *Lectures on exponential decay of solutions of second order elliptic operators*. Princeton Univ. Press.
Allegretto, W. (1974) On the equivalence of two types of oscillation for elliptic operators. *Pacific J. Math.* **55**, 319–28.
Ancona, A. (1986) On strong barriers and an inequality of Hardy for domains in $\mathbb{R}^n$. *J. London Math. Soc.* (2) **34**, 274–90.
Ancona, A. (1987) Negatively curved manifolds, elliptic operators, and the Martin boundary. *Ann. Math.* **125**, 495–536.
Anderson, M., Schoen, R. (1985) Positive harmonic functions on complete manifolds of negative curvature. *Ann. Math.* **121**, 429–61.
Anker, J.-P., Lohoué, N. (1986) Multiplicateurs sur certains espaces symetriques. *Amer. J. Math.* **180**, 1303–54.
Aronson, D. G. (1968) Non-negative solutions of linear parabolic equations. *Ann. Sci. Norm. Sup. Pisa* (3) **22**, 607–94.
Azencott, R. (1974) Behaviour of diffusion semi-groups at infinity. *Bull. Soc. Math. France* **102**, 193–240.
Bakry, D., Emery, M. (1984) Hypercontractivité de semi-groupes des diffusion. *C. R. Acad. Sci. Paris* t. 299 Serie 1, 775–7.
Bakry, D., Emery, M. (1985) *Diffusions hypercontractives*. Sem. Prob. XIX. Lecture Notes in Math. Vol. 1123, 177–206. Springer-Verlag.
Baouendi, M. S., Goulaouic, C. (1969) Régularité et théorie spectrale pour une classe d'operateurs elliptique dégénérés. *Arch. Rat. Mech. Anal.* **34**, 361–79.
Beardon, A. F. (1983) *The geometry of discrete groups*. Graduate texts in Math. Vol. 91 Springer-Verlag.
Berard, P. H. (1986) *Spectral geometry: direct and inverse problems*. Lecture Notes in Math., Vol. 1207. Springer-Verlag.
Berg, M. van den (1984A) A uniform bound for trace ($e^{t\Delta}$) for convex regions in $\mathbb{R}^n$ with smooth boundaries. *Comm. Math. Phys.* **92**, 525–30.
Berg, M. van den (1984B) On the spectrum of the Dirichlet Laplacian for horn-shaped regions in $\mathbb{R}^n$ with infinite volume. *J. Funct. Anal.* **58**, 150–6.
Berg, M. van den (1987A) On the asymptotics of the heat equation and bounds on traces associated with the Dirichlet Laplacian. *J. Funct. Anal.* **71**, 279–93.
Berg, M. van den (1987B) Bounds for the Dirichlet heat kernel. Preprint.
Bergh, J., Löfström, J. (1976) *Interpolation spaces: an introduction*. Springer-Verlag.
Birman, M. S., Solomjak, M. Z. (1980) Quantitative analysis in Sobolev imbedding theorems and applications to spectral theory. *Amer. Math. Soc. Transl. Sec. 2*, **114**, 118–23.
Bougerol, Ph. (1983) Exemples de théorèmes locaux sur les groupes resolubles. *Ann. Inst. H. Poincaré* **29B**, 369–91.

References

Brooks, R. (1981A) A relation between growth and the spectrum of the Laplacian. *Math. Zeit.* **178**, 501–8.
Brooks, R. (1981B) The fundamental group and the spectrum of the Laplacian. *Comment. Math. Helv.* **56**, 581–98.
Brooks, R. (1985) The bottom of the spectrum of a Riemannian covering. *Crelle's Journal* **357**, 101–14.
Calabi, E. (1965) An extension of E. Hopf's maximum principle with an application to geometry. *Duke Math. J.* **25**, 285–303.
Carlen, E. A., Kusuoka, S., Stroock, D. W. (1987) Upper bounds for symmetric Markov transition functions. *Ann. Inst. H. Poincaré* **23**, 245–87.
Carmona, R. (1974) Regularity properties of Schrödinger and Dirichlet semigroups. *J. Funct. Anal.* **17**, 227–37.
Cassels, J. W. S. (1986) *Local fields*. London Math. Soc. Student Texts. Vol. 3, Cambridge Univ. Press.
Chavel, I. (1984) *Eigenvalues in Riemannian geometry*. Academic Press.
Cheeger, J., Gromov, M., Taylor, M. (1982) Finite propagation speed, kernel estimates for functions of the Laplace operator, and the geometry of complete Riemannian manifolds. *J. Diff. Geom.* **17**, 15–53.
Cheeger, J., Taylor, M. (1982) Diffraction of waves by conical singularities. *Comm. Pure Appl. Math.* **35**, 275–331, 487–529.
Cheeger, J., Yau, S. T. (1981) A lower bound for the heat kernel. *Comm. Pure Appl. Math.* **34**, 465–80.
Chernoff, P. (1973) Essential self-adjointness of powers of generators of hyperbolic equations. *J. Funct. Anal.* **12**, 401–14.
Cordes, H. O. (1987) *Spectral theory of linear differential operators and comparison algebras*. London Math. Soc. Lecture Notes Vol. 76, Cambridge Univ. Press.
Courant, R., Hilbert, D. (1953) *Methods of mathematical physics*. Vol. 1. Interscience.
Courant, R., Hilbert, D. (1966) *Methods of mathematical physics*. Vol. 2. *Partial differential equations*. Interscience.
Croke, C. B., Derdzinski, A. (1987) A lower bound for λ_1 on manifolds with boundary. *Comment. Math. Helv.* **62**, 106–21.
Cycon, H. L., Froese, R. G., Kirsch, W., Simon, B. (1987) *Schrödinger operators, with applications to quantum mechanics and global geometry*. Springer-Verlag.
Davies, E. B. (1976) *Quantum theory of open systems*. Academic Press.
Davies, E. B. (1980) *One-parameter semigroups*. Academic Press.
Davies, E. B. (1982) JWKB and related bounds on Schrödinger eigenfunctions. *Bull. London Math. Soc.* **14**, 273–84.
Davies, E. B. (1983) Hypercontractive and related bounds for double well Schrödinger operators. *Quart. J. Math. Oxford Ser.* (2) **34**, 407–21.
Davies, E. B. (1984) Some norm bounds and quadratic form inequalities for Schrödinger operators, II. *J. Oper. Theory* **12**, 177–96.
Davies, E. B. (1985A) Trace properties of the Dirichlet Laplacian. *Math. Zeit.* **188**, 245–51.
Davies, E. B. (1985B) The eigenvalue distribution of degenerate Dirichlet forms. *J. Diff. Eqns.* **60**, 103–30.
Davies, E. B. (1985C) Criteria for ultracontractivity. *Ann. Inst. H. Poincaré* **43A**, 181–94.
Davies, E. B. (1985D) L^1 properties of second order elliptic operators. *Bull. London Math. Soc.* **17**, 417–36.
Davies, E. B. (1986) Perturbations of ultracontractive semigroups. *Quart. J. Math. Oxford* (2) **37**, 167–76.
Davies, E. B. (1987A) Heat kernel bounds for second order elliptic operators on Riemannian manifolds. *Amer. J. Math.* **109**, 545–70.

References

Davies, E. B. (1987B) Explicit constants for Gaussian upper bounds on heat kernels. *Amer. J. Math.* **109**, 319–34.

Davies, E. B. (1987C) The equivalence of certain heat kernel and Green function bounds. *J. Funct. Anal.* **71**, 88–103.

Davies, E. B. (1988A) Kernel estimates for functions of second order elliptic operators. *Quart. J. Math. Oxford* (2) **39**, 37–46.

Davies, E. B. (1988B) Gaussian upper bounds for the heat kernels of some second order operators on Riemannian manifolds. *J. Funct. Anal.*

Davies, E. B., Mandouvalos, N. (1987) Heat kernel bounds on manifolds with cusps. *J. Funct. Anal.* **75**, 311–22.

Davies, E. B., Mandouvalos, N. (1988) Heat kernel bounds on hyperbolic space and Kleinian groups. *Proc. London Math. Soc.*

Davies, E. B., Simon, B. (1984) Ultracontractivity and the heat kernel for Schrödinger operators and Dirichlet Laplacians. *J. Funct. Anal.* **59**, 335–95.

Davies, E. B., Simon, B. (1986) L^1 properties of intrinsic Schrödinger semigroups. *J. Funct. Anal.* **65**, 126–46.

Davies, E. B., Simon, B., Taylor, M. (1988) L^p spectral theory of Kleinian groups. *J. Funct. Anal.* **78**, 116–136.

Debiard, A., Gaveau, B., Mazet, E. (1976) Théorèmes de comparaison en géometrie Riemannienne. *Research Inst. Math. Science, Kyoto Univ.* **12**, 391–425.

Dodziuk, J., Kendall, W. S. (1986) Combinatorial Laplacians and isoperimetric inequality. *Proc. Warwick Symp. on Stochastic differential equations and applications*, ed. K. D. Elworthy. *Pitman Research Notes in Math.*, Vol. 150, pp. 68–74. Pitman.

Donnelly, H. (1986) Bounded harmonic functions and positive Ricci curvature. *Math. Z.* **191**, 559–65.

Donnelly, H., Li, P. (1982) Lower bounds for the eigenvalues of Riemannian manifolds. *Mich. Math. J.* **29**, 149–61.

Doyle, P. G. (1988) On the bass note of a Schottky group. *Acta Math.*

Dunford, N., Schwartz, J. T. (1958) *Linear operators*, Part 1. Interscience.

Duren, P. L. (1983) *Univalent functions*. Springer-Verlag.

Eckmann, J. P. (1974) Hypercontractivity for anharmonic oscillators. *J. Funct. Anal.* **16**, 388–404.

Edmunds, D. E., Evans, W. D. (1987) *Spectral theory and differential operators*. Clarendon Press.

Elstrodt, J. (1974) Die Resolvente zum Eigenwertproblem der automorphen Formen in der hyperbolischen Ebene III. *Math. Ann.* **203**, 99–132.

Fabes, E. B., Stroock, D. W. (1986) A new proof of Moser's parabolic Harnack inequality via the old ideas of Nash. *Arch. Rat. Mech. Anal.* **96**, 327–38.

Faris, W. (1978) Inequalities and uncertainty principles. *J. Math. Phys.* **19**, 461–66.

Fefferman, C., Sánchez-Calle, S. (1986) Fundamental solutions for second order subelliptic operators. *Ann. Math.* **124**, 247–72.

Fleckinger, J. (1981) *Asymptotic distribution of eigenvalues of elliptic operators on unbounded domains*. Lecture Notes in Math. Volume 846, 119–28. Springer-Verlag.

Frehse, J. (1977) Essential self-adjointness of singular elliptic operators. *Bol. Soc. Bras. Mat.* **8.2**, 87–107.

Fukushima, M. (1977) On an L^p estimate of resolvents of Markov processes. *Research Inst. Math. Science, Kyoto Univ.* **13**, 277–284.

Fukushima, M. (1980) *Dirichlet forms and Markov processes*. North-Holland.

Gilbarg, D., Trundinger, N. S. (1977) *Elliptic partial differential equations of second order*. Springer-Verlag.

Grisvard, P. (1981/2) Singularities for the problem of limits in polyhedra. *Seminaire Goul. Schwartz*, Ex. No. 8.

Gross, L. (1976) Logarithmic Sobolev inequalities. *Amer. J. Math.* **97**, 1061–83.
Hardy, G. H., Littlewood, J. E., Polya, G. (1952) *Inequalities.* 2nd Edition. Cambridge Univ. Press.
Hasminskii, R. Z. (1960) Ergodic properties of recurrent diffusion processes and stabilisation of the solution of the Cauchy problem for parabolic equations. *Theor. Prob. and Appl.* **5**, 179–96.
Hayman, W. K. (1977/8) Some bounds for principal frequency. *Applicable Anal.* **7**, 247–54.
Hempel, R., Voigt, J. (1986) The spectrum of a Schrödinger operator in $L_p(R^v)$ is p-independent. *Comm. Math. Phys.* **104**, 243–50.
Hempel, R., Voigt, J. (1987) On the L^p spectrum of Schrödinger operators. *J. Math. Anal. Appl.* **121**, 138–59.
Hille, E. (1962) *Analytic function theory*, Vol. 2. Ginn & Co.
Hueber H. (1985) *A uniform estimate for Green functions on $C^{1,1}$ domains.* BiBoS preprint.
Hueber, H., Sieveking, M. (1982) Uniform bounds for quotients of Green functions on $C^{1,1}$ domains. *Ann. Inst. Fourier* **32**, 1, 105–17.
Jerison, D. S., Sánchez-Calle, A. (1986) Estimates for the heat kernel for a sum of squares of vector fields. *Indiana Univ. Math. J.* **35**, 835–54.
Kadlec, J., Kufner, A. (1966) Characterisation of functions with zero traces by integrals with weight functions. *Časopis Pest. Mat.* **91**, 463–71.
Kalf, H., Walter, J. (1972) strongly singular potentials and essential self-adjointness of singular elliptic operators in $C_0^\infty(\mathbb{R}^n\setminus\{0\})$. *J. Funct. Anal.* **10**, 114–30.
Kato, T. (1978) Trotter's product formula for an arbitrary pair of self-adjoint contraction semigroups. Pp. 185–95 of *Topics in Functional Analysis*, eds. I. Gohberg and M. Kac. Academic Press.
Kato, T. (1981) Remarks on the self-adjointness and related problems for differential operators. *Spectral theory of differential operators, Proc. Conf. Univ. Alabama.* Pp. 253–66. North-Holland.
Kaufman, R. P., Wu, J.-M. (1982) Parabolic potential theory. *J. Diff. Eqns.* **43**, 204–34.
Kifer, Y. (1986) Brownian motion and positive harmonic functions on complete manifolds of non-positive curvature. *Proc. Warwick Symp. on stochastic differential equations and applications*, ed. K. D. Elworthy. *Pitman Research Notes in Math.* Vol. 150, pp. 187–232. Pitman
Koranyi, A., Taylor, J. C. (1983) Fine convergence and parabolic convergence for the Helmholtz equation and the heat equation. *Ill. J. Math.* **27**, 77–93.
Koranyi, A., Taylor, J. C. (1985) Minimal solutions of the heat equation and uniqueness of the positive Cauchy problem on homogeneous spaces. *Proc. Amer. Math. Soc.* **94**, 273–8.
Kusuoka, S., Stroock, D. (1987) Applications of Malliavin calculus, Part. 3. *J. Fac. Sci. Tokyo Univ., Sect. 1A, Math.* **34**, 391–442.
Kusuoka, S., Stroock, D. (1988) Long time estimates for the heat kernel associated with a uniformly subelliptic second order operators. *Ann. Math.* **127**, 165–89.
Lax, P. D., Phillips, R. S. (1982) The asymptotic distribution of lattice points in Euclidean and non-Euclidean spaces. *J. Funct. Anal.* **46**, 280–350.
Lax, P. D., Phillips, R. S. (1985) Translation representations for automorphic solutions of the wave equation in non-Euclidean spaces III. *Comm. Pure Appl. Math.* **38**, 179–207.
Léandre, R. (1987A) Majoration en temps petit de la densité d'une diffusion dégénérée. *Prob. Th. Rel. Fields* **74**, 289–294.
Léandre, R. (1987B) Minoration en temps petit de la densité d'une diffusion dégénérée. *J. Funct. Anal.* **74**, 399–414.
Li, P. (1986) Large time behaviour of the heat equation on complete manifolds with non-negative Ricci curvature. *Ann. Math.* **124**, 1–21.
Li, P., Tam, L.-F. (1987) Positive harmonic functions on complete manifolds with non-negative curvature outside a compact set. *Ann. Math.* **125**, 171–207.

Li, P., Tam, L.-F. (1988) Symmetric Green's functions on complete manifolds. *Amer. J. Math.* to appear.
Li, P., Yau, S. T. (1980) Estimates of eigenvalues of a compact Riemannian manifold. *Proc. Symp. Pure Math.* **36**, 205–239.
Li, P., Yau, S. T. (1986) On the parabolic kernel of the Schrödinger operator. *Acta Math.* **156**, 153–201.
Lichnerowicz, A. (1958) Geometrie des groupes des transformations. Dunod, Paris.
Lieb, E. H. (1983) on the lowest eigenvalue of the Laplacian for the intersection of two domains. *Inv. Math.* **74**, 441–8.
Lions, J. L., Magenes, E. (1972) *Non-homogeneous boundary-value problems and applications*, Vol. 1. Springer-Verlag.
Lohoué, N. Rychener, Th. (1978) *Some function spaces on symmetric spaces related to convolution operators*. Preprint.
Lohoué, N., Rychener, Th. (1982) Resolvente von Δ auf symmetrischen Räumen vom nichtkompakten Typ. *Comment. Math. Helv.* **57**, 445–68.
Lyons, T. (1987) Instability of the Liouville property for quasi-isometric Riemannian manifolds and reversible Markov chains. *J. Diff. Geom.* **26**, 33–66.
Lyons, T. (1988) *On Makaroo's law of the iterated logarithm*. Preprint.
Lyons, T., Sullivan, D. (1984) Function theory, random paths and covering spaces. *J. Diff. Geom.* **19**, 299–323.
Mair, B., Taylor, J. C. (1984) Integral representation of positive solutions of the heat equation. *Lecture Notes in Math.* **1096**, 419–33, Springer-Verlag.
Mandouvalos, N. (1983) *The theory of Eisenstein series and spectral theory for Kleinian groups*. Cambridge Univ. Ph.D. thesis.
Mandouvalos, N. (1986) The theory of Eisenstein series for Kleinian groups. *Contemporary Math.* **53**, 357–70.
Mandouvalos, N. (1987) Scattering operator, Eisenstein series, inner product formula and 'Maass–Selberg' relations for Kleinian groups. *Mem. Amer. Math. Soc.*, to appear.
Mandouvalos, N. (1988) Spectral theory and Eisenstein series for Kleinian groups. *Proc. London Math. Soc.*, to appear.
Maz'ja, V. G. (1979) *Einbettungssätze für Sobolevsche Räume*, Vol. 1, *Teubner Text für Mathematik*. Teubner.
McKean, H. P., Singer, I. M. (1967) Curvature and the eigenvalues of the Laplacian. *J. Diff. Geom.* **1**, 43–69.
Miller, K. (1967) Barriers on cones for uniformly elliptic operators. *Ann. Mat. Pura Appl.* **76**, 93–105.
Miller, K. (1971) Extremal barriers on cones with Phragmén-Lindelöf theorems and other applications. *Ann. Mat. Pura Appl.* **90**, 297–329.
Molchanov, A. M. (1953) On conditions for discreteness of the spectrum of self-adjoint differential equations of second order. *Ttrudy Mosk. Mat. Obs.* **2**, 169–200.
Moser, J. (1961) On Harnack's theorem for elliptic differential equations. *Comm. Pure Appl. Math.* **14**, 577–91.
Moser, J. (1964) A Harnack inequality for parabolic differential equations. *Comm. Pure Appl. Math.* **17**, 101–34.
Moser, J. (1967) Correction to 'A Harnack inequality for parabolic differential equations'. *Comm. Pure Appl. Math.* **20**, 232–6.
Moss, W., Piepenbrink, J. (1978) Positive solutions of elliptic equations. *Pacific J. Math.* **75**, 219–26.
Mueller, C., Weissler, F. (1982) Hypercontractivity for the heat semigroup for ultraspheric polynomials and on the n-sphere. *J. Funct. Anal.* **48**, 252–83.
Nagel, A., Stein, E. M., Wainger, S. (1985) Balls and metrics defined by vector fields, I. Basic properties. *Acta Math.* **155**, 103–47.
Nash, J. (1958) Continuity of solutions of parabolic and elliptic equations. *Amer. J. Math.* **80**, 931–54.

Nelson, E. (1966) A quartic interaction in two dimensions. *Mathematical theory of elementary particles.* Pp. 69-73, MIT Press.
Nelson, E. (1973) The free Markov field. *J. Funct. Anal.* **12**, 211-27.
Oddson, J. (1978) On the boundary point principle for elliptic equations in the plane. *Bull. Amer. Math. Soc.* **74**, 666-70.
Osserman, R. (1977) A note on Hayman's theorem on the bass note of a drum. *Comment. Math. Helv.* **52**, 545-55.
Osserman, R. (1980) Isoperimetric inequalities and eigenvalues of the Laplacian. *Proc. Inter. Cong. Math. (Helsinki 1978)*, Acad. Sci. Fennica, Helsinki, 435-42.
Pang, M. M. H. (1987) L^1 properties of singular second order elliptic operators. *J. London Math. Soc.* to appear.
Patterson, S. J. (1976) The limit set of a Fuchsian group. *Acta Math.* **136**, 241-73.
Patterson, S. J. (1983) Further remarks on the exponent of convergence of Poincaré series. *Tohoku Math. J.* **35**, 357-73.
Patterson, S. J. (1987) Lectures on measures on limit sets of Kleinian groups. *Analytical and geometric aspects of hyperbolic space*, ed. D. B. A. Epstein, London *Math. Soc. Lecture Notes*, Vol. 111, pp. 281-323, Cambridge Univ. Press.
Perry, P. (1987A) The Laplace operator on a hyperbolic manifold I. Spectral and scattering theory. *J. Funct. Anal.* **75**, 161-87.
Perry, P. (1987B) The Laplace operator on a hyperbolic manifold, II. Eisenstein series and the scattering matrix for certain Kleinian groups. Preprint.
Persson, A. (1964) Compact linear mappings between interpolation spaces. *Ark. Mat.* **5**, 215-19.
Phillips, R. S., Sarnak, P. (1985A) On the spectrum of the Hecke groups. *Duke Math. J.* **52**, 211-21.
Phillips, R. S., Sarnak, P. (1985B) The Laplacian for domains in hyperbolic space and limit sets of Kleinian groups. *Acta Math.* **155**, 173-241.
Pinchover, Y. (1986) Sur les solutions positives d'équations elliptiques et paraboliques dans $\mathbb{R}^n$ *Comptes Rendus Acad. Sci. Paris* Série 1, **302**, 447-50.
Pinchover, Y. (1987) *Representation theorems for positive solutions of parabolic equations.* Preprint.
Porper, F. O., Eidel'man, S. D. (1984) Two sided estimates of fundamental solutions of second order parabolic equations and some applications. *Russian Math. Surveys* **39** (3), 119-78.
Reed, M., Simon, B. (1975) *Methods of modern mathematical physics.* Vol. 2. *Fourier analysis, self-adjointness.* Academic Press.
Reed, M., Simon, B. (1978) *Methods of modern mathematical physics.* Vol. 4. *Analysis of operators.* Academic Press.
Röckner, M., Wielens, N. (1985) Dirichlet forms - closability and change of speed measure. Pp. 119-144 of *Infinite dimensional analysis and stochastic processes*, ed. S. Albeverio. Research Notes in Mathematics, Vol. 124. Pitman.
Rogers, L. C. G., Williams, D. (1986) Construction and approximation of transition matrix functions. Pp. 133-66 of Special Suppl. to *Adv. in Appl. Prob.*, papers in honour of G. E. H. Reuter. Publ. by Appl. Prob. Trust and London Math. Soc.
Rosen, J. (1976) Sobolev inequalities for weight spaces and supercontractivity. *Trans. Amer. Math. Soc.* **222**, 367-76.
Rosenbljum, V. G. (1972) The eigenvalues of the first boundary value problem in unbounded domains. *Math. USSR - Sb.* **18**, 235-48.
Rosenbljum, V. G. (1973) The calculation of the spectral asymptotics for the Laplace operator in domains of infinite measure, *Problems of Mathematical Analysis*, 4. Pp. 95-106, Izdat Leningrad, Univ. Leningrad.
Rothaus, O. S. (1981) Diffusion on compact Riemannian manifolds and logarithmic

Sobolev inequalities. *J. Funct. Anal.* **42**, 102–9.
Rothaus, O. S. (1985) Analytic inequalities, isoperimetric inequalities and logarithmic Sobolev inequalities. *J. Funct. Anal.* **64**, 296–313.
Rothaus, O. S. (1986) Hypercontractivity and the Bakry-Emry criterion for compact Lie groups. *J. Funct. Anal.* **65**, 358–67.
Sánchez-Calle, A. (1984) Fundamental solutions and geometry of the sum of squares of vector fields. *Inv. Math.* **78**, 143–60.
Silverstein, M. L. (1974) *Symmetric Markov processes. Lecture Notes in Math.* Vol. 426, Springer-Verlag.
Simon, B. (1974) *The $P(\phi)_2$ Euclidean quantum field theory*. Princeton Univ. Press.
Simon, B. (1979) *Functional integration and quantum physics*. Academic Press.
Simon, B. (1982) Schrödinger semigroups. *Bull. Amer. Math. Soc.* **7**, 447–526.
Simon, B. (1983A) Some quantum operators with discrete spectrum but classically continuous spectrum. *Ann. Phys.* **146**, 209–20.
Simon, B. (1983B) Non-classical eigenvalue asymptotics. *J. Funct. Anal.* **53**, 84–98.
Stein, E. M. (1970) *Singular integrals and differentiability properties of functions*. Princeton Univ. Press.
Stein, E. M., Weiss, G. (1971) *Introduction to Fourier analysis on Euclidean spaces*. Princeton Univ. Press.
Strichartz, R. (1983) Analysis of the Laplacian on the complete Riemannian manifold. *J. Funct. Anal.* **52**, 48–79.
Stroock, D. W. (1984) *An introduction to the theory of large deviations*. Springer-Verlag.
Stroock, D. W., Varadhan, S. R. S. (1979) *Multidimensional diffusion processes*. Springer-Verlag.
Sullivan, D. (1979) The density at infinity of a discrete group of hyperbolic motions. *Publ. Math. IHES* **50**, 171–202.
Sullivan, D. (1984) Entropy, Hausdorff measures old and new, and limit sets of geometrically finite Kleinian groups. *Acta Math.* **153**, 259–77.
Sullivan, D. (1987) Related aspects of positivity in Riemannian geometry. *J. Diff. Geom.* **25**, 327–51.
Tamura, H. (1976) The asymptotic distribution of eigenvalues of the Laplace operator in an unbounded domain. *Nagoya Math. J.* **60**, 7–33.
Terras, A. (1985) *Harmonic analysis on symmetric spaces and applications*, I. Springer-Verlag.
Triebel, H. (1978) *Interpolation theory, function spaces, differential operators*. North-Holland.
Varadhan, S. R. S. (1967) Diffusion processes in small time intervals. *Comm. Pure Appl. Math.* **20**, 659–85.
Varopoulos, N. Th. (1981) The Poisson kernel on positively curved manifolds. *J. Funct. Anal.* **44**, 359–80.
Varopoulos, N. Th. (1982) Green's functions on positively curved manifolds, II. *J. Funct. Anal.* **49**, 170–6.
Varopoulos, N. Th. (1983) Potential theory and diffusion on Riemannian manifolds. Pp. 821–37 of *Proc. Conf. on harmonic analysis in honor of A. Zygmund*. Chicago, eds. W. Beckner *et al.* Wadsworth.
Varopoulos, N. Th. (1984) Brownian motion and random walks on manifolds. *Ann. Inst. Fourier* **34**, 243–69.
Varopoulos, N. Th. (1985A) A potential theoretic property of groups. *Bull. Sci. Math.* **109**, 113–19.
Varopoulos, N. Th. (1985B) Isoperimetric inequalities and Markov chains. *J. Funct. Anal.* **63**, 215–39.
Varopoulos, N. Th. (1985C) Hardy–Littlewood theory for semigroups. *J. Funct. Anal.* **63**, 240–60.

Varopoulos, N. Th. (1985D) Long range estimates for Markov chains. *Bull. Sci. Math.* **109**, 225–52.

Varopoulos, N. Th. (1985–6) *Analyse sur les groupes de Lie. Première partie: groupes nilpotents.* Lecture notes, Universite de Paris VI.

Varopoulos, N. Th. (1986) Théorie du potentiel sur des groupes et des variétés. *C. R. Acad. Sci. Paris*, Ser. 1 **302**, 203–5.

Vulis, I. L., Solomjak, M. Z. (1972) Spectral asymptotics of degenerate elliptic operators. *Soviet Math. Doklady* **13**, 1484–8.

Weyl, H. (1912) Das asymptotische Verteilungsgesetz der Eigenwerte linearer partieller Differentialgleichungen. *Math. Ann.* **71**, 441–69.

Yau, S. T. (1975) Harmonic functions on complete Riemannian manifolds. *Comm. Pure Appl. Math.* **28**, 201–28.

Zhao, Z. X. (1986) Green function for Schrödinger operator and conditioned Feynman-Kac guage. *J. Math. Anal. Appl.* **116**, 309–34.

Notation index

L^p, L^∞	1
$\wedge$	1
$\vee$	1
f_+, f_-	1
$\lvert f \rvert$	1
C_c^∞	4
$\mathscr{S}$	4
$\hat{f}$	4
Q	7
$\mathrm{Quad}(H)$	7
$\mathscr{B}_c$	8
$\mathscr{B}_0$	8
$\mathscr{B}_{\mathrm{loc}}$	8
$W^{1,p}(\Omega)$	8
H_p	23
$d(x)$	27
Inr	28
tr	52
$\partial\Omega$	107
∇, Δ	147
$d(x,y)$	147
Ric	153
$\sigma(x,y)$	176
$\delta(\Gamma)$	187

Index

Banach lattice 1
Beurling–Deny conditions 12
Bochner–Lichnerowicz–Weitzenbock formula 184
bottom of the spectrum 5, 156
bounded geometry 167, 172
bounded holomorphic semigroup 23

Calabi's theorem 153
compact resolvent 39
compactness and spectrum 35
complete Riemannian manifold 151, 153, 157
complex times 102
conformal metric 135, 148
conservation of probability 24, 153
contraction semigroup 14, 22
convex polyhedron 181
convolution 3
cotangent space 146
cusp 181
cut locus 153

difference operator 19
diffusion 179
Dirichlet boundary conditions 10
Dirichlet form 22, 148
  classification 56
discrete spectrum 5
distance function 147
divergence 147

elliptic operator 10
ergodic theory 181
essential self-adjointness 56, 151
essential spectrum 5
Euclidean algorithm 180
exponent of convergence 182
exponential volume growth 156
extension property 46

external ball condition 27
external cone condition 27, 130

Feller property 155
Feynman–Kac formula 52
first passage time 166
form limit 8
Fourier transform 2
Friedrichs extension 11
functions of an elliptic operator 99
fundamental domain 181
fundamental point pair invariant 177

Gaussian lower bound 82, 91
Gaussian upper bound 82, 83
geodesic 177
Golden–Thompson inequality 53
gradient 147
graph Laplacian 20
Green function 4, 83
ground state 109
group
  amenable 157
  Fuchsian 180
  fundamental 180
  geometrically finite 182
  Hecke 181
  Kleinian 180
  rank one 177
  Schottky 182

Hadamard's method of descent 178
Hardy's inequality 25
harmonic function 34
  bounded 165, 184
  positive 165, 184
harmonic oscillator 113
Harnack inequality 21, 98, 157
Hausdorff dimension 183
Hausdorff–Young inequality 2

Index

heat kernel 4, 59
Holder's inequality 2
Hopf boundary point lemma 126
Hunt process 56
hyperbolic ball 176
hyperbolic space 176
hypercontractive semigroup 59, 80, 114
hypoelliptic operator 90, 106

inradius 28
internal cone condition 129
invariant set 24
irreducible semigroup 24
isometry group 157, 177

Koebe's one-quarter theorem 32

Laplacian 4, 147
Lichnerowicz theorem 166
limit set 183
Lipschitz boundary condition 27
Lipschitz function 6
logarithmic Sobolev inequality 65
logarithmically convex 141

Markinciewicz interpolation theorem 76
Markov chain 19
Markov process 21
Martin boundary 166
Mehler's formula 113
minimal solution 166
minimax 5
mollifier 9
Moser's Harnack inequality 82

Nash inequalities 78
Neumann boundary conditions 11
normal local coordinates 158
null set 1
number theory 181

orbifold 181
ordinary set 183

parabolic Harnack inequality 157
Patterson measure 183
Phrágmen–Lindelöf theorem 104
Poisson summation formula 54
potential 49, 119
    local singularities 51, 138
potential theory 164
pseudo-resolvent 37

quadratic form 6
    closure 7
    core 7
    non-negative 7
quantum field theory 59, 80

random walk 19
reflection principle 107
region
    bounded 125
    conical 128
    convex 43, 47
    polygonal 127
    regular 27
    strongly regular 28
regularised distance 6
Ricci curvature 153, 157
Riemannian distance 147, 176
Riemannian metric 146
Riesz–Thorin interpolation theorem 3
Rosen's lemma 117

scattering theory 181
Schottky domain 182
Schrödinger operator 48, 119
Schwartz space 4
Selberg's theorem 180
singular elliptic operator 131
Sobolev inequality 40, 75
Sobolev space 8
spectrum, L^p 61, 115, 178
Stein interpolation theorem 3
strictly elliptic operator 10
subexponential volume growth 156
subharmonic comparison theorem 118
subharmonic function 34
superharmonic function 34
surface measure 26
symmetric Markov semigroup 21

tangent space 146
trace class 52
transient manifold 157, 176
Trotter product formula 49

ultracontractive semigroup 59, 80
uniformly elliptic operator 10
uniformly hypoelliptic operator 82

vector field 146

weighted L^2 space 109
Weyl's theorem 52
Whitney's theorem 6

Young's inequality 3

Printed in the United States
By Bookmasters